WILD
WATERS

A wildlife and water lover's
companion to the aquatic world

WILD WATERS

A wildlife and water lover's
companion to the aquatic world

Susanne Masters

Illustrated by Alice Goodridge

Vertebrate Publishing, Sheffield
www.v-publishing.co.uk

WILD WATERS
Susanne Masters
Illustrated by Alice Goodridge

First published in 2021 by Vertebrate Publishing.

VERTEBRATE PUBLISHING
Omega Court, 352 Cemetery Road, Sheffield S11 8FT, United Kingdom.
www.v-publishing.co.uk

This book is a work of non-fiction. The author has stated to the publishers that, except in such minor respects not affecting the substantial accuracy of the work, the contents of the book are true.

A CIP catalogue record for this book is available from the British Library.

ISBN: 978-1-83981-100-5 (Paperback)
ISBN: 978-1-83981-101-2 (Ebook)

10 9 8 7 6 5 4 3 2 1

Design and production by Jane Beagley, Vertebrate Publishing.
www.v-publishing.co.uk

Vertebrate Publishing is committed to printing on paper from sustainable sources.

Printed and bound in Europe by Latitude Press.

SAFETY STATEMENT

Being in and around water and interacting with wildlife carries a risk of personal injury or death. People must be aware of and accept that these risks are present and they should be responsible for their own actions and involvement. Nobody involved in the writing and production of this book accepts any responsibility for any errors that it may contain, nor are they liable for any injuries or damage that may arise from its use.

Pay attention to currents, tides and weather conditions, read safety signs, know your limits and be vigilant at all times. Take care near the edges of banks and in boggy areas. Some wildlife can be harmful to humans and pets. If necessary, seek medical advice. If foraging, seek permission if required, be certain of what you are picking, do not collect rare species, take no more than you plan to consume and only collect from plentiful populations.

Many animals, plants and habitats are protected by legislation, including the Wildlife and Countryside Act 1981. Be careful not to disturb wildlife at any time. Keep pets under control, especially at vulnerable times of year for wildlife, such as nesting, pupping or spawning season. Aim to leave no trace of your visit. Act responsibly to enjoy and preserve all that the aquatic environment has to offer.

Contents

Introduction

I've learnt about aquatic life, in part, through dipping into presentations at conservation science conferences and reading updates from assorted IUCN Species Survival Commission groups. Ostensibly I've been at these conferences and receive these communications because of my research on wild orchid harvesting. This dry and cerebral approach has been an adjunct to reality. Like everyone, scientist or not, I have my own relationship with nature that isn't confined to reading about it or discussing it. And my interest in the living world isn't restricted to orchids – a group of plants that occasionally favour boggy places but never get fully immersed. My journeys in, on and around water are a source of experience and also questions that send me back to reading and learning about what I've seen.

I was fished out of a couple of ponds as a kid who fell into them while looking at their contents. This is unexceptional; most kids are thrilled at the realms contained within ponds, and do need supervision around bodies of water. Childhood summers in Switzerland with the Swiss side of my family were mostly occupied with swimming in various lakes. Alongside my mum (who taught me to swim), dad and grannies, who were happy to spend time in and next to the water, my sister was a good swimmer. Probably I was mostly an annoying younger sister, but we did swim off to the floating pontoons and explore lake edges together.

Life got a little wilder when my family moved from a town in the south-east to a rural house on the Suffolk–Essex border. A mysterious puff of fluff skittering around the yard turned out to be a moorhen chick brought back to the house by my cat. Jeremy, who lived above the stables, helped to catch it and we returned the chick to Long Pond. Mrs Duck nested in the peonies each spring and launched her ducklings in our garden pond before marching them off along the footpath towards wilder waters. A visiting city friend was amazed to see our neighbour Ben practising his fly-fishing cast in the meadow by his house. Sometimes Ben and I would take our dogs down to the gravel pits for walks. Filled with water, the pits were a lush oasis in hot East Anglian summers. I have some of Ben's watercolours in my house now – landscapes from his Scottish fishing trips, Long Pond seen from the footpath, and a speckled trout.

Going to university and then working in cities, I swam in swimming pools indoors, and enjoyed wild water while on holiday abroad. This drought was broken by a couple of seaside camping weekends, followed by a campervan trip with Lesley and Biffer for eight months. Mostly we traced the coast of England, Wales and then Scotland, with the addition of a few forays inland. Biffer participated with collie-dog paranoia in canoe and boat rides, but full-heartedly jumped into all opportunities to be in water. While I greatly appreciate and value explorations of far-flung places, finding eight months' travel insufficient to explore England, Wales and Scotland made it clear that delight and excitement isn't tied to the exotic.

When we temporarily lived in Purbeck, Biffer was my first swimming buddy. His company gave me confidence to venture on early morning swims with no one else around. He trotted along the shore and swam after his tennis ball, which I threw while swimming. Settling in Bournemouth, a seaside town, daily walks to the beach for swim-loving Biffer showed how even on a repeated route, water and wildlife creates variety. Winter storms cast lobster and by-the-wind sailors on the beach; in summer, vast barrel jellyfish washed ashore.

In a town by the sea there are plenty of swimmers, many of whom became friends. At some point, Sally and a few others started sending me photos asking questions about animals and plants they were seeing on the way to swims and when they were in the water. I pestered Jonathan, the editor of *Outdoor Swimmer*, about writing on wildlife from the perspective of people encountering it while in the water, to answer the kinds of questions that I was being asked, from 'What is this?' to 'Will it sting or bite me?' – meeting wildlife in its element carries additional vulnerability. That wildlife column grew into the Eco Pages, and it's fantastic that Jonathan and the editorial team have created space for wildlife every month for years.

Alice Goodridge's illustrations were instrumental in showing Jonathan how the water-wildlife column could work, and it's been a special privilege to go on to make this book with her as a friend who also happens to be a talented illustrator. We first met at the beach in Bournemouth, years before working together. Since then, we've seen a lot of wildlife and wild places, including swimming over kelp beds from a boat to St Kilda's shore, and dodging jellyfish on a Dublin beach.

Alongside Alice, microadventures with Lucy, Keith and Amanda have ranged through swimming among waterlilies to warming up post-swim with hot cocktails flavoured with waterside plants. They've shaped great memories and spawned a stack of maps I've annotated with swimming spots, perils such as quicksand warnings, and wild camping places. (Thanks to Alastair Humphreys for clarifying the concept of and offering the word 'microadventure'.)

At home, bioluminescent swims with Gary have elevated average working days and stretched the feeling of summer into autumn dusks and dawns. Sam has shown me a different way to enjoy the sea, and should be the number one jellyfish fan that I know – unless that number one fan is Gaynor, who is working on cracking the secret of embalming a dead jellyfish. In contrast, Holly and Patrick share jellyfish horror, which I don't feel, even though I've been stung a few times. It's valuable to be reminded that not everyone feels the same. Similarly, Annette was a good sport when I laughed that she was terrorised by a shoal of tiny fish and also tolerant of my sending them her way to demonstrate the amazing high-speed coordination with which they avoid crashing into each other or people in the sea. With Ilse, it's always precious to see how the wildlife we've met on swims takes form in her art. Gemma is lovely company on swims and foraging, and her enthusiasm for learning about wildlife is contagious. Ali and Marc have

improved my ability to travel distance with their swimming tips, and Ali has also been the tide guru for working out when tidal stretches of rivers are swimmable.

Perhaps my niece's perspective is skewed, but her faux horror at a demonstration of seaweed edibility while we were still in the sea and thrilled delight at spotting a large pike watching us in a river is an endorsement of real-world pleasures surpassing the thrall of electronic spaces. Amy has been up for adding swims to science trips, and related adventures such as forage-from-a-punt-without-using-hands; the speed and height of a punt is reasonable for biting off edible leaves and flowers in passing. Myles is a long-term enabler of finding and spending time in and on water, including road-tripping the whole of America's east coast and working from a park bench while I swam in a Tennessee waterfall pool. He instigated paddleboarding, which was a soothing escape from a hot summer that also carried the sorrow of Biffer dying. In that strange pandemic winter, my dad was as keen as I was to make repeated visits to a winterbourne. Over months, seeing the watercourse greening with watercress, trickling with water between patches of mud, and then flowing clear and fast, was a reassuring connection with place and season. I must also give kudos and thanks to Cassy and Maria, who despite the warm comfort of their Florida and California origins ventured beyond our Texas and Florida swimscapades to embrace the cold and have joined loch swims to Scottish castles and been buffeted by chilly English waves.

Through friends, stories of wildlife and stories of people become intertwined. Keith being rammed by a territorial salmon; Kate up for trying out a sugar-kelp foot rub and taking time to watch iridescent sea gooseberries and comb jellyfish. Looking back into history, we can see how human culture grows from biodiversity. Wildlife isn't a museum piece. Wild species are part of our heritage and still resources that we rely on.

Wildlife can be harmed by our presence – for example, disturbing birds from hunting for their food or trampling vegetation on riverbanks. The more informed we are about how wild animals and plants live and grow, the better equipped we are to navigate enjoying being outdoors and the wildlife we find there while avoiding harming it. When people move gently through wild spaces with eyes open, they can also be an ally noticing when wildlife and habitats are harmed. Wildlife crime takes place unseen; police cannot continuously survey rural areas. Harm to wildlife can be reported to wildlife crime officers. Noticing and holding companies accountable for pollution events is a vital step towards reducing their occurrence. Even small acts such as disentangling broken fishing line from waterweeds and putting it in a bin benefit wild spaces and the life that inhabits them.

This book is a companion written for myself in the past when I was beginning to explore water, and for other people venturing waterside in their mind or with footsteps who are curious about how we are connected to aquatic life. It is a skin of pages, words and pictures that draws together a multitude of influences and stories. Like a map, this book is a tool for situating yourself within the wildlife that surrounds you, and for moving towards where you want to go.

Clearly this introduction is also acknowledgement of the people who have given me encouragement and companionship in enjoying wild waters and the life that inhabits it – map makers, book writers, friends and family. And a hope for those reading that they also find encouraging companions in the form of book, hound or person.

1

Immersed
on Land

**Attention-grabbing rainforests are not
the sole source of the oxygen our lives
depend on. Every other breath is filled
with oxygen provided by dinoflagellates,
ocean-dwelling microscopic plants.
On land we remain immersed.**

If you open your fridge or kitchen
cupboards, aquatic plants and animals are
staple foods and essential components of
our food chains. Some life-threatening
diseases are treated and cured with the
chemical arsenal possessed by life that
inhabits water. This isn't remote and exotic
wildlife we can only meet in distant places
or see on TV. If you walk out of your door
along canal paths and seashores, or stop
to look in a murky pond in a park, you will
find wildlife that affects our lives.

We are taught that water and land are
separate. Maps show clear boundary lines
along riverbanks, lakeshores and coasts, but
life is messy and sprawling. Coastal cliffs
are clad in seaweeds at their depth, then
salt-tolerating lichen above the high-tide
mark, followed by seashore plants above the
splash zone. Seabirds fish the open ocean
and nest on rocky ledges. Our separation of
land and water is more driven by our own
limitation – we breathe air, and only visit
water. Instead, think of water like a forest:
rivers woven into land like roots of trees and
reaching out into a canopy of sea. Even on
dry land with no view of water, we remain
connected to the life aquatic in all its forms,
from oceanic to riparian.[1]

< Wildlife that savours reed beds (*see page 176*)

---

1 *Riparian* is derived from the Latin *riparius*, meaning 'bank'
or 'shore'.

Meadowsweet (*Filipendula ulmaria*)

In sickness and health

When we reach for medicine at home, it comes out of a packet: pills popped from blister packs, ointment out of a tube, a teaspoonful of liquid poured from a bottle. Similarly, in hospital settings, pills, drips and injections are ways of administering drugs. It is all too easy to feel alienated by thinking chemical cocktails are foreign to us. We used to rely on wildlife around us to take care of our health, and even now our medicinal toolkit contains many treatments that are derived from or were inspired by natural sources. On the shores of Lough Neagh there aren't many people left who will wrap an eel-skin bandage around a

sprain, even though eel-skin bandages were a traditional Ulster remedy and there are still plenty of eels in the Lough. Now most people with a sprain will reach for aspirin – one of our most commonly used pharmaceuticals – for pain relief. It is derived from water-loving trees.

Seeing greyish-green leaves in the distance, the fluff of white seeds blowing in a breeze or a lime-green weeping tree are all long-distance sightings of willow trees – telling you that water is near. Willow trees have eased our pain for more than 3,500 years. In 1862 Edwin Smith, an American living in Cairo in Egypt, bought some ancient scrolls in Luxor that dated back to 1500 BC. One of the remedies written across them was on the use of willow to treat pain. Before this Egyptian evidence was revealed, willow was known as a herbal remedy in Europe. In 1763 Reverend Edward Stone wrote about using willow bark to treat fever. In 1828 Johann Buchner discovered that salicin is the active ingredient in willow. His discovery of the chemical basis to willow's efficacy is the reason we don't resort to raw willow bark for pain relief now. We still use this chemical to treat pain in a refined form as aspirin. Further scientific endeavour led to Charles Gerhardt's modification of salicin to a concentrated form called salicylic acid. In 1897 Felix Hoffmann used salicylic acid from meadowsweet (*Filipendula ulmaria*) – another riparian plant – to create acetylsalicylic acid, the stable form of aspirin we use as a painkiller today. Aspirin's essential raw ingredient, salicylic acid, is now commercially manufactured with a chain of reactions starting with phenol and hydroxide, reacted with carbon dioxide and then acidified. Yet without millennia of willow bark being used successfully as a traditional remedy, and help from meadowsweet in isolating the essential compound, we would not have aspirin in our pharmacy cabinets today. Pharmaceuticals regulate dosage so that each pill or tablet contains a consistent

amount of active ingredients. Aspirin's living origins of willow and meadowsweet on green riverbanks are not just a metaphor for calm but also a literal source of soothing for modern times.

In Britain and Ireland, white willow (*Salix alba*) – most likely the species of willow whose use was recorded by the Egyptians – is an introduced species that we see most often near water. Other species include crack willow (*Salix fragilis*), named for its easily and noisily broken branches, and osier (*Salix viminalis*). Willow trees will sweep low over water, providing a convenient temporary mooring post if you are paddling or swimming. Perhaps the finest species to pause next to is in gardens and parks where weeping willows, a cultivated form (*Salix babylonica*), are planted. Their domed interior is hidden from view and bathed in light greenly filtered through leaves.

Contemporary medical science still relies on a basic piece of equipment supplied by the sea. Agar is a seaweed product that is a critical tool for biomedical science. It creates a jelly that is perfect for scientists to cultivate bacteria and fungi, enabling the identification of diseases and testing of potential cures. Agar is clear, which makes it easy to see experimental results. Powdered agar is added to hot water and nutrients required for microbial growth. While hot, it remains liquid and can be poured to fill petri dishes. Once cold, it sets and provides a solid surface on which to observe experiments.

Gelidium is the genus of seaweed that is most used to make laboratory-grade agar. Its naturally high gelling strength combined with the low temperature at which it gels are desirable characteristics not found in agar made from other genera of seaweed. *Gelidium* agar is in high demand from biotechnology, biomedicine and pharmacology.

Agar contains agarose and agaropectin. Some experiments involving DNA analysis demand pure agarose. Agar also contains sulphates, which are negatively charged and can change the way DNA moves through agar. On the other hand, agarose is neutral – without positive or negative charge – and this allows DNA and RNA's movement through agarose to be analysed. Using an electric current to propel DNA through agarose, molecules are sorted by their size and electrical charge. This is how genetic fingerprints, identification of individuals via their unique DNA profile, are possible. Perhaps more famous from crime-scene analysis, genetic fingerprinting has medical applications such as diagnosing inherited disease. Running RNA amplified by polymerase chain reaction (PCR) through agarose is also a way to detect the presence of viruses – such as coronavirus – in people.

It provides a means to directly test whether people are infected, rather than looking for indirect markers like immune response.

Scientists around the world panicked in 2015 when Morocco, the country that harvested eighty-two per cent of the world's *Gelidium*, restricted export. The price of agar tripled. Suppliers of laboratory-grade agar products reduced the range of different blends and concentrations of agar that they offered. Laboratories rationed their stocks of agar.

Moroccan *Gelidium* had become dominant because of its cheap cost supported by low wages, and because seaweed hand-picked by snorkellers and divers was of a better quality than the beach-collected *Gelidium* supplied by other countries. Morocco was seeking to increase the economic benefit of this harvest by extracting agar in-country and being paid for the more valuable derivative rather than the cheap raw material. This was accompanied by a significant decrease in yield – there was less *Gelidium* to pick. Harvesting *Gelidium* washed up on beaches was sustainable. But picking fresh *Gelidium* without restricting its harvest had become unsustainable.

Gelidium seaweeds grow on rocks in cool water that are washed by turbulence. This keeps them supplied with nutrients and keeps them within their growing temperature. These conditions are difficult to replicate under cultivation.

Realising the vulnerability of a critical laboratory supply coming from one place has galvanised attempts to cultivate *Gelidium* and interest in harvesting it from other countries. Production of laboratory-grade agar from *Gracilaria* seaweeds is also increasing.

Like many seaweeds, *Gelidium* is on the move. Changing sea temperatures are pushing it out of some areas and welcoming it into others. Its travel is supported by boat traffic moving it to new locations. It has started to appear around the coast of Britain and Ireland – not yet in harvesting abundance, but it's an interesting seaweed to spot if you consider that there are very few people who have never had cause to thank an agar plate's work in taking care of their own health or that of a loved one. Agar plates are still used in identifying infectious diseases and remain a critical tool in the process of developing anti-microbial drugs.

Agar doesn't work well for tissue culture of animal cells – a vital means of modelling animal responses outside of animal bodies. Instead, collagen has been used to create a jelly matrix for growing animal cells. Tissue culture has reduced the use of animals in biomedical testing, and is essential for its contribution to understanding cancer cells, pharmaceutical toxicity testing and production of vaccines.

Animal tissue culture has relied on collagen derived from cows and pigs, but now jellyfish are being used as an alternative that has the advantage of not carrying diseases such as bovine spongiform encephalopathy (BSE). Barrel jellyfish (*Rhizostoma pulmo*) are the easiest jellyfish to spot when you are on or in the sea. They are also the main species of jellyfish used as a source of collagen because they are large, abundant and easy to catch.

Oceans and seas aren't just frontiers where wildlife is still being discovered, or a source for tools used by medical science; they are one of the biggest sources of novel pharmaceuticals. Marine bioprospecting – searching the sea for natural sources of molecules and discovering biochemical and genetic information – is at the forefront of developing new medicine. Historically, our use of marine wildlife for medicinal treatments has been restrained by how inhospitable marine environments are for humans. Technological advances in diving equipment for people and making remote-controlled machines able to operate underwater have provided a pathway. Impetus has been given by the realisation that marine life produces a range of chemicals that simply aren't found on land, and this offers a particular advantage in seeking defences from one of the biggest threats to global human health – antibiotic resistance.

Since Alexander Fleming discovered penicillin in 1928, we've lived in a golden age in which survival rates from surgery and recovery from illnesses have been significantly improved by mass-produced effective antibiotics preventing bacterial infection. It is an era that has come to an end with recent cases of people dying from bacteria that were resistant to all available antibiotics. Pneumonia, tuberculosis, blood poisoning, sexually transmitted diseases, post-operative infections and foodborne diseases are all becoming harder to treat.

There are some compounds present in marine life that terrestrial bacteria haven't been exposed to, and these offer an advantage in the arms race between people and bacteria on land. Harpoon weed (*Asparagopsis armata*), a fluffy pink seaweed that was first noted in English scientific records in Australia and Aotearoa (New Zealand) in 1855 might offer us a weapon. It is a seaweed of the Pacific and Indian Oceans that has moved into the Mediterranean and Atlantic. First seen in Galway Bay in 1939 and in the Bristol Channel in 1949, it has recently been the target of assessment for antibacterial activity. It was found to inhibit growth in a range of bacteria, making it one of the wild species offering hope in the hunt for new antibiotics.

Foods

Beyond food such as fish and shellfish that we directly recognise as being harvested from aquatic environments, watery wildlife permeates our food production system as it transfers nutrients and other useful compounds from water to land.

Harpoon weed – as well as having potential in our antibiotic armoury – might help to reduce the environmental impact of food production. Cows are notorious for farting out methane and accelerating climate change. In fairness, climate change can't entirely be blamed on cows, and pasture-fed cattle are essential elements in maintaining some of our rare meadow wildlife through grazing. One target in identifying ways to have beef, milk and leather with a lower environmental footprint has been reducing the amount of methane that cows produce. Including harpoon weed in cattle feed has been found to reduce methane production by up to sixty-seven per cent. It isn't unnatural for cattle to eat seaweed. In a few locations around Ireland and Britain, herds with coastal grazing wander on the beach and eat a little seaweed along with other food they find there.

Lamb that has grazed on salt marsh is sold for a premium. It is thought the salt-marsh plants flavour the meat of the sheep which eat them. Salt-marsh lamb has ancient roots. Written records document expanses of salt marsh used for grazing in medieval England. Buried evidence uncovered by archaeologists shows that in the Bronze Age, people relying on domestic grazing animals made use of salt marsh around the Severn Estuary area of England as productive land on which to feed their animals.

Lamb isn't the only product that attracts a premium when it is nurtured by coastal plants; the world's most expensive potatoes are fuelled by seaweed. On the French island of Noirmoutier, new potatoes are harvested young and sold as 'La Bonnotte'. Along with sandy soil and excellent marketing, these potatoes are shaped by using seaweed as a fertiliser. Jersey potatoes have never reached the peak of £400 per kilo that La Bonnotte potatoes have achieved, but Jersey is also an island that markets its potatoes as seaweed-fertilised. Alongside the nutrients that seaweed contains, there is an advantage in comparison to compost – seaweed contains no agricultural weed seeds. Seventeenth- and nineteenth-century records of penalties and fines imposed in Jersey include matters concerning the collection of *vraic*, as seaweed was locally named. Seaweed featuring in the legal system indicates how valuable it was to people living on Jersey.

Landlocked gardeners who don't have a nearby seashore for collecting seaweed often use seaweed as a fertiliser for tomatoes without realising – since one of the most popular tomato fertilisers sold in shops uses liquid seaweed extract as a key ingredient. Seaweeds are also showing promise as a way to boost plant health, thereby reducing the economic impacts of plant diseases such as blight. Improving plants' resistance to disease is appealing as a sustainable means of looking after crops without applying chemicals that are toxic not just to pathogens but also to ecosystems by leaving residue in soil and on crops and killing insects.

Another marine-derived fertiliser is fish blood and bone, which supplies the three main nutrients that plants need to grow: nitrogen, phosphorous and potassium. An advantage this organic fertiliser has over chemical extracts is that it yields these nutrients slowly. Slow-release fertilisers need less frequent application and they don't leach so many nutrients into waterways as there is more time for plants to absorb them.

Clean wild water is essential for a couple of water-loving plants that are cultivated for mass supply. Watercress (*Nasturtium officinale*) was such an important crop grown in Hampshire's chalk streams that 'the Watercress Line', a railway track, was built to connect Alresford's watercress crops with the mainline train to London. Watercress grows best in clear-running cool water, just as it wells up in the springs that feed Hampshire's chalk streams. In order to make use of wild water to grow watercress for sale, people built flat watercress beds filled by natural streams that then flowed back into their natural course. While the Watercress Line now operates as a novelty ride, Hampshire watercress is still for sale across the country – although roads provide quicker and cheaper transport now. Britain's love for watercress isn't sated by homegrown produce; it is also shipped in from Spain. In contrast to the soft lush countryside of Hampshire, a lot of Spanish watercress is grown on the rather arid and mountainous

islands of Tenerife and Gran Canaria, in the Canary Islands. A shared feature between these two different landscapes is the man-made beds through which naturally fresh and clean spring water runs, creating the habitat that watercress thrives in.

Watercress and wasabi (*Eutrema japonicum*) are relatives, both being members of the cabbage family (*Brassicaceae*). Wasabi brings heat to Japanese cuisine, but it grows wild along cool mountain streams. It's difficult to grow wasabi because of the conditions it requires, which are not dissimilar to the water quality that watercress demands. England's commercially grown wasabi crop

Watercress (*Nasturtium officinale*)

is nurtured in old watercress beds on the Dorset–Hampshire border.

Waterside places are also home to the wild relatives of a couple of the world's major crops. In early summer, large stalks of lemon yellow flowers opening above grey cabbage leaves along the seashore mark out an otherwise indistinct plant. Most people walk past wild cabbage (*Brassica oleracea*), the wild plant that spawned so many of our vegetables. Cabbage, cauliflower and broccoli are fairly ubiquitous across all the continents. Brussels sprouts and kohlrabi are more regionally known. Gai lan and collard greens are respectively Chinese and American specialities. Cavolo nero is a kind of kale with Italian heritage. Green kale was a stalwart vegetable in British gardens during the Second World War. All of these vegetables are cultivars of sea cabbage. One of the benefits that sea cabbage passes on to these crops is tolerance of low water availability – since seaside plants often have to cope with limited freshwater. Crops that don't need daily watering are better equipped to tolerate low-maintenance care.

Another seaside plant is the ancestor of one of our most colourful vegetables and also had a role in the abolitionist movement. Sea beet (*Beta vulgaris*) is a shiny-leaved seashore plant. Beetroots and sugar beet are selections of this plant that develop large sweet roots. At the base of the leafstalks on sea beet, you can often find fine lines of the bright vermilion colour we associate with beetroots, although wild sea beet does not have the swollen roots of garden cultivars of beetroot. Sugar beet was developed as a crop in Silesia, a region that included areas that are now in Poland, the Czech Republic and Germany.

In Georgian England, consumers selected sugar produced in India by 'free people' as an alternative to sugar provided by the transatlantic slave trade. In America, without the colonial resource of India to supply sugar, beet sugar was promoted by The Beet Sugar Society of Philadelphia as sugar produced without the abuse of slaves, although it wasn't until after the Civil War that sugar-beet cultivation became viable in the USA.

In England, the importance of sugar beet as a crop led to an emergency authorisation for use of a product containing thiamethoxam, a neonicotinoid, to be used in 2021, even though its use outdoors was prohibited in 2018. A cold winter killed enough aphids, which are carriers of beet yellows virus, for the virus forecast to be below the threshold set by the emergency authorisation for use of thiamethoxam. But as our winters are no longer reliably cold, these aphids and the virus they carry may yet again raise conflict between economics and environmental protection. It will take several years to develop sugar beet with better natural resistance to the aphids that this previously banned chemical works on as a pesticide. But it's possible that help in developing sugar beet that better resists these aphids might be found by reintroducing genes from its wild crop ancestor – sea beet.

Material matters

Every taste of Scotch whisky relies on a wild river plant. In whisky, as much as grain, water and peat shape the taste, it is the casks that transform raw spirit. Distilleries use imported bourbon and sherry barrels to impart elements of additional flavour into their whisky. Before being used, casks are reassembled and checked for watertightness. Bulrush (*Schoenoplectus lacustris*) is used to create a seal between the ends and sides of the barrels.

Tall shiny green cylinders of bulrush on a river can reach above head height. When passing by, it is worth pausing to snap a stem off and take a look at the secret ingredient that helps to make whisky. On the outside, it is waxy and bright green. After peeling off the green skin, its interior is a surprise – white and looking very much like synthetic sponge. When these stalks are dried, they are ideal as long, flexible lengths of foam filler that aren't toxic – a natural sealant.

Peat makes its flavour known in whisky, but it is so strong that only small amounts are used in the distilling industry. Peat bogs have been denuded by horticultural demand for peat as a growing medium. Its ability to hold water and nutrients is considered ideal, yet the vast scale of demand caused the deplorable destruction of an important wild habitat. Peat is an ancient vintage of sphagnum moss, in which hundreds and thousands of years of growth have gradually been compressed.

Around the world there are nearly 400 species of *Sphagnum*. Amidst this diversity, they share the characteristic that, dead or alive, they store large volumes of water within their structure. Even in living sphagnum moss, ninety per cent of the cells are dead yet still serve the function of being available to hold water. Sphagnum mosses grow while saturated with water, and as they accumulate they can extend their wet habitat. Because of the electrical charge produced in its cell walls, sphagnum releases positively charged ions that make the water in its surrounding environment acidic, which inhibits bacterial growth. People took advantage of this prior to the advent of electrical freezing and the use of chemical adulterants in food preservation by making bog butter. Animal fat placed in wooden containers was buried in bogs. Bogs are slow to warm up – moderated by the mass of water they hold – so this kept bog butter cool. The acidic and low-oxygen conditions also limited microbial growth.

The preservative power of peat bogs has provided insight into past lives. Most human remains quickly reduce to bones. In the embrace of cool, low-oxygen and acidic bog water, bodies decay slowly. Bodies retrieved from bogs in various degrees of mummification can sometimes have their skin and internal organs preserved. In 2011 the Cashel Man was found in Cul na Móna bog in County Laois, Ireland. He died around 2000 BC and is the oldest bog body found intact. His injuries, accompanied by his body being found with wooden stakes and having been intentionally covered with peat, suggest that his death was not an unwitnessed misadventure in the bog. The Cashel Man was ritually killed. His discovery at a geographically significant location and death in an era in which nobles' lives were offered when their people suffered hardship such as bad harvests or sickness makes it likely that he was a nobility sacrifice.

During the First World War, sphagnum was a life-saving natural resource. With thousands of people injured, supplies of clean bandages were running out. Old stories about the Battle of Clontarf in 1014 reported that moss was used to pack the wounds of the injured. A botanist and a military surgeon collaborated and identified two species of *Sphagnum* that worked to staunch wounds and promote healing – *S. papillosum* and *S. palustre*. By 1918 – with British harvests of sphagnum supplemented by supplies collected in British Columbia and Nova Scotia by the Canadian Red Cross – every month a million wound dressings were being sent from Britain to hospitals in Europe, Egypt and Western Asia.

A different aquatic plant that is still used as a material is in plain view – common reed (*Phragmites australis*). Sending up fast-growing new growth each spring, common reed forms thickets along slow-moving stretches of fresh or brackish water. Dead stalks from previous years' growth persist. Although old stems dry to beige sticks, they remain upright unless cut or knocked and are slow to decay. This makes them a useful material for roof-thatching, and as they grow in dense stands it is efficient to harvest them in bulk. Since cutting the reeds doesn't kill the plant, it is a renewing resource that can be repeatedly harvested.

Reed beds also form a habitat that is home to some of Britain's rare species. Bearded tits (*Panurus biarmicus*), little birds whose males have dark feathers leading off their beaks like handlebar moustaches, are only found in reed beds. Although plant diversity in reed beds is low – being dominated by the reeds themselves – a special plant lives amongst them. Milk parsley (*Peucedanum palustre*) is the only food plant that caterpillars of swallowtail butterflies (*Papilio machaon britannicus*) will eat. It is the presence of milk parsley amongst reeds in the Norfolk Broads that makes this the place to see Britain's biggest butterfly.

Marks on the landscape

In and on water, aquatic wildlife is part of the scenery. Its realm of influence also extends beyond aquatic habitats into shaping our physical and cultural landscape. Ancient maps had monsters included on them. A lot of them weren't species we recognise today, but at the time they were considered real. In the seventeenth century, animals on maps were more an illustration of where resources could be found than areas to avoid because of dangerous beasts.

Across the world, human settlements have orientated around freshwater resources. Rivers in particular were desirable not just for supply of water for household and agricultural use, but also for the transport and trade opportunities enabled by river travel. The construction of aqueducts was an early innovation that enabled populations to sprawl beyond drinking-water sources. However, the introduction in the industrial era of equipment to pump groundwater, and desalination techniques that make salty water potable, have loosened the connection with freshwater. Coastal areas, being a liminal space, offered people food resources from both land and sea; in particular, shorelines provided rich rewards to gatherers.

Sometimes on seashores, stones seem to have fallen in lines that don't follow organic curves but are crudely straight. From close up, with some stones out of line and a coating of barnacles and seaweed, it can be hard to realise what you are seeing. Looking down from a loftier viewpoint on a clifftop, it becomes clear that these V-shaped lines of rocks pointing away from land are not natural features. They are man-made rock pools that take advantage of the difference in water levels between high and low tides. It takes a lot of work to move stones and build fish traps. But once constructed, checking the landside of these simple stone traps is a low-effort means of collecting fish that had been swimming in the area behind the trap and were caught as the tide fell.

Poppit Sands in Pembrokeshire kept a secret from hordes of visitors, until an image taken by Google Earth on a sunny day showed a dark 'V' standing out against an otherwise turquoise backdrop of clear water against pale sand. Over time, its stones have sunk into the sand and are no longer uncovered at low tide. But at the time of its construction, thought to be around 1066, it would have been exposed at low tides and provided its builders with fresh fish. At Strangford Lough in County Down, the monks of Grey Abbey are thought to have been the builders of the now partially dismantled lines of stones in the sand. Around Britain and Ireland's shores, fish traps built in stone were an important supplier of food.

England recorded ownership of land and resources in the Domesday Book, an extensive survey compiled in 1086. Its intended use was in administration and law. In hindsight, the Domesday Book also tells us about the social structure of England and what natural resources were considered assets. Because estate holders were asked how many fisheries they had,

there is a large amount of information on where fish were abundant enough to be regularly harvested. Fisheries include shellfish. So we know that Whitstable Oyster beds in Kent were being used as a regular source of oysters at that time. Salmon (*Salmo salar*) records occurred more often in western districts. Lampreys (*Lampetra*) were at that time a popular food: one manor in Surrey recorded 1,000 of them. This was forty-nine years before King Henry I was reported as dying from eating too many lampreys, although his death is now thought to have been caused by food poisoning rather than fish.

In medieval Europe, money was not the only currency – eels (*Anguilla anguilla*) were legal tender. People could pay rent to

European eel (*Anguilla anguilla*)

their landlords in sticks of eels. One stick was twenty-five eels. For example, in 1086 William of Warenne was owed 20,000 eels. And in 1547 Thomas Boteler leased two water mills to Thomas Sankey for twenty-one years for a rent of six pounds, thirteen shillings and four pence plus 300 sticks of eels.

Eels were in demand as a food for everyone from peasants to kings. But demand may have exceeded supply. From the fourteenth century onwards, fewer rents were collected in eels. Archaeological records contain smaller eel remains, a sign that eels were not attaining the larger sizes associated with longer lifespan before death. England looked abroad to sustain its appetite for eels.

In the fourteenth century, the Dutch exported salted eels to England. By the fifteenth century, technical advances made it possible for them to trade live eels with England. Eel ships filled their holds with water and live eels. These ships were a regular feature of the Thames, and as such were marked on maps showing them at their regular docks in Queenhithe and Billingsgate. Business was disrupted by the Anglo–Dutch wars, but afterwards returned to London markets, and the trade in live eels continued until 1938.

Throughout the era of Dutch imports, eel catching continued in England. In 1820 an Eel House was built on the River Alre in Hampshire. When the sluices were opened, river water ran through water channels in the building but iron grilles caught eels. The riverkeeper would transfer the live eels to boxes that he towed downstream to his cottage. Merchants collected the eels and took them to fish markets, including London with its high concentration of the population. Centuries of demand for eels is etched on British history within financial documents, coats of arms, maps and even buildings created especially for them.

At our westernmost edge, clinging to life on sparse rocky islands amidst a million seabirds, the people of St Kilda survived by eating seabirds and their eggs and paying rent to their landlord in feathers and bird oil. On the harbour, the feather house is still standing – empty of feathers now that they are no longer in demand for stuffing pillows and bedding, but a monument to an isolated community who survived by making use of birds. After 2,000 years of human occupation, St Kilda became uninhabited. Its last residents were evacuated in 1930 as life there had become untenable. People still visit the island, no longer out of curiosity about 'the bird people', but rather to see the birds and the abandoned village that rests in the custody of the National Trust for Scotland's rangers.

Ballan wrasse (*Labrus bergylta*)

Surprising values

Ballan wrasse (*Labrus bergylta*) are common in rocky coasts around Britain and Ireland. Not sought after as a fish for eating, and too small to offer epic 'sport' in a tussle to reel them in on a line, they drew little attention until in the 1990s someone realised that ballan wrasse could be used to address one of the problems that plagues salmon farms: sea lice. These parasites thrive in salmon populations crowded into sea pens, and develop resistance to the chemical treatments released to kill them. Ballan wrasse and several other closely related species will pick sea lice off salmon. This organic solution to the problem of sea lice masks an unsustainable wild harvest – it relies on taking ballan wrasse from the sea, shipping them to salmon pens and then leaving them captive. With their price peaking at £50,000 per tonne, ballan wrasse were named the UK's most expensive fish. This drew attention to

their plight, and led to the introduction of catch limits and, in some areas, bans on exporting wrasse. Ultimately, wrasse will probably remain victims of hunger for farmed salmon unless a kinder solution is found for managing sea lice or consumers refuse to buy salmon produced using the captive labour of ballan wrasse that should have been left wild and free.

Some of the most valuable fish in the world were seen on a flying visit to England's coast in 2020. A kayaker spotted Atlantic bluefin tuna (*Thunnus thynnus*) in a feeding frenzy offshore from Plymouth. A group of tuna had herded a shoal of smaller fish towards the surface, and as the tuna launched themselves at their prey at high speed, sometimes they popped out above the sea. At about two metres long, these are unmistakable fish, and their flesh is so delicious that it commands high prices. In 2019, one bluefin tuna was sold for £2.5 million in Japan. There are three species of bluefin tuna, all of which are edible, scarce and impossible to breed in captivity. Their flavour coupled with their rarity makes them expensive to buy. Around the UK it is illegal to catch tuna. Angling for Atlantic bluefin tuna in Irish waters is allowed under a byelaw introduced in 2020 for authorised boats. Given the speed and power that they use in their predatory lifestyle, fishing for bluefin tuna is sometimes likened to big game hunting.

In 1967 the River Tay yielded one of the world's most famous natural pearls when a diver found the Abernethy pearl. It is remarkably spherical for a wild pearl. Notably large, its size indicates that the freshwater pearl mussel (*Margaritifera margaritifera*) it grew in was old enough to become large itself. These mussels have been known to live for up to 130 years. The International Union for Conservation of Nature (IUCN) has listed freshwater pearl mussels as at risk of extinction. Within their distribution around northern Europe, north-eastern Canada and the USA, Britain and Ireland are the global stronghold. Populations of freshwater pearl mussels were reduced by centuries of pearl collection in which hundreds and sometimes thousands of them were collected and killed in searches that might yield just one pearl. Post-industrial-era water pollution was also catastrophic for colonies of these molluscs

that need fast-running clear water to live in. In 1998 it became illegal to collect these mussels, but the survival of the species has still been perilous.

Freshwater pearl mussels yield valuable pearls, and they have had a considerable impact on legislation protecting wildlife in Scotland. In 2013 a firm was prosecuted for pollution offences that included killing freshwater pearl mussels, but was only fined £4,000. It was felt that this was an inadequate penalty for destroying an entire colony of a globally rare species. Prompted by this and concerns raised by other cases, Scotland undertook a public review of wildlife crimes and penalties imposed. Wildlife in Scotland now has fiercer protection in terms of the sentences that can be passed. Currently the fine that can be imposed for killing, injuring or taking a pearl mussel is £5,000 or six months in jail. These penalties are applicable for each freshwater pearl mussel.

Taken out of its natural environment, some of our wildlife has considerable worth that has shaped history and can drive unsustainable harvests. Many of these species do not have a secure future. However, legislative changes prompted and supported by public opinion can help – and are helping – to retain these species as part of our living natural heritage and ensure that they have a wild future.

Painful encounters

Sometimes time spent in and around water can leave people with painful reminders. Despite anxiety fuelled by cinematic portrayal of sharks and the presence of about forty species of sharks passing through or living in British and Irish waters, shark bites are incredibly unlikely. Our only records of shark bites around Britain and Ireland involve sharks that were being handled after having been caught on fishing line or nets. Rather than being anxious about fears of monsters from the deep, once equipped with knowledge of anything that might result in a painful encounter, you can enjoy wildlife without undue worry.

While watching animals and birds, it is always important to avoid cornering them. It's not just a case of thinking about where they are in relation to you and fixed objects like rocks; it's also taking into account the proximity of other people and how collectively you may be preventing wildlife from getting away. Leaving wildlife without a clear escape route pushes it to try and force a way past you. Animals that are habituated to humans might be easy to watch, but be wary. These are animals without fear of people, and are more likely to be close enough to bite or peck you. If you feel compelled to rescue wildlife that looks like it is in trouble, remember that your intentions are not known to the animal. Already stressed, they are primed to act defensively – it is better to leave animal rescues to the numerous organisations that take care of our wildlife, or at least get advice from them before attempting to make an intervention.

While we are primed to be alert to danger from obvious and large creatures, most painful encounters with aquatic wildlife around Britain and Ireland are a surprise. It is the stealthy and unobtrusive wildlife that gets the chance to make contact, leaving us with a painful memento.

One of the key threats from wild water is so small that you need a microscope to see its structure. Blue–green algae are cyanobacteria that naturally occur in freshwater and produce chemicals harmful to both people and animals. Warm weather in summer is when conditions encourage their populations to multiply and they can reach harmful concentrations. Although you can't see the chemicals they produce, clusters of these bacteria become big enough to be visible as turquoise or green scum or flecks of colour on the surface of water. For people, contact with blue–green algae and their toxic chemicals can cause skin rashes and eye irritation, vomiting, fever, diarrhoea, muscle pain and joint pain. Animals can die. Protect yourself and your pets by checking water before getting in and if it looks scummy, cancel being in the water or find a different body of water. Dogs

that swim in water containing blue–green algae and lick their wet fur afterwards, or drink from it, are vulnerable to illness with sudden onset and must be taken to a vet.

Filamentous algae and duckweed (*Lemna*) can be confused with blue–green algae. In the case of filamentous algae, if you poke it with a stick you should find that it is composed of green hair-like strands. Examining it more closely, if what looks like green scum is in fact a raft of tiny plants with miniature leaves, it is duckweed. If blue–green algae floating on surface water is disturbed, it breaks into small particles or smaller clumps. The Centre for Ecology & Hydrology provides a phone app called Bloomin' Algae, through which reports of blue–green algae are uploaded and displayed on a map. You can use it to check locations to see if blue–green algae have been reported there.

Forensic mapping identified how a painful new arrival to Britain and Ireland was travelling and spreading. By marking reports of giant hogweed (*Heracleum mantegazzianum*) on maps, it became clear that it was spreading along riverbanks. When fully grown, it towers above head height and is topped by a parasol of white flowers. It can take several years to flower, but at all stages of its life cycle, touching it can cause large blisters.

< Sharks that inhabit British and Irish seas (*see page 177*)

Giant hogweed was brought into Britain and Ireland as an ornamental plant – if you see one flowering you will understand its aesthetic appeal. Many garden plants are kept for ornamental purposes and their violent touch is tolerated and ignored, which can be manageable within a contained area. But giant hogweed escaped from gardens. In 1828 it was found growing wild in Cambridgeshire – a fast migration to the wild, considering that it was first recorded in England in 1817 on the seed list of the Royal Botanic Gardens in Kew. It is now listed across Europe as a species of concern. Giant hogweed's method of attack contains a sneaky time delay. You can brush against it and feel nothing. Then fifteen minutes later your skin is inflamed to the extent that painful marks looking like burns can arise. Indeed, they are burns. Giant hogweed's sap contains furanocoumarins, which are phytotoxic and prevent your skin from being able to protect itself from sunlight. Contact with giant hogweed leaves you with amplified vulnerability to sunlight. If you do belatedly realise that you have touched a plant, wash the affected area with cold water and soap and stay out of sunlight for forty-eight hours. If you realise that you made contact with giant hogweed too late to avoid damage, and develop phytophotodermatitis – plant- and light-induced skin inflammation

– seek medical treatment for the pain and to reduce the chance of infection in damaged skin.

Not only plants but also animals can sting. While mature jellyfish can give you an unexpected sting if you don't spot them, a fair amount of the time they can be avoided by swimming around them. Occasionally a jellyfish bloom can concentrate them into unavoidable numbers – but you can see them before you get into the sea and wisely choose to swim elsewhere or at another time.

It is juvenile jellyfish that are the most irritating. If you swim wearing anything that lets a little water in, baby jellyfish can land in there. While water can pass through most fabric, jellyfish won't. As you swim along, your swimwear sieves the sea and keeps hold of those little jellyfish. A sting from one immature jellyfish has negligible impact. But there is going to be more than one caught in there. You might not notice until you are out of the water and red welts or itchy patches flare up in areas covered by your swimwear. Eventually the itchiness will pass and your skin should go back to normal. Peak season for this to happen is late spring and early summer. As the year progresses, jellyfish get bigger and won't get swept between your swimwear and your skin.

In most cases when you see wildlife there is far more risk that you will harm it than that it will harm you. Getting to know sensitive times of year and places where different species are particularly vulnerable is a good way of reducing your impact on wildlife you want to enjoy. Some wildlife is protected by legislation. Other wildlife relies on your goodwill.

There is a rich history of wildlife in and around water shaping our life on land. Our relationship with aquatic life continues with new uses of wildlife being discovered, and is enhanced by developing sustainable strategies for making wild harvests.

Jellyfish found around Britain and Ireland (*see page 176*) >

2

Inhabited

Life isn't spread homogenously across our planet. Even within the realm of wildlife associated with water, a range of influences – geology, topography, salinity and human activity – combine and create habitats that some but not all species can live in. The wildlife we see tells us about these influences that are often invisible to the passing or naked eye. Wildlife is a cast of characters that informs us about places we are in.

The River Thames illustrates how habitats change over the course of a river, and under different influences. It is hard to imagine the clear-running stream slipping between trees in a Gloucestershire meadow as the same entity as the broad and brackish expanse of water between the Houses of

Parliament and the South Bank. In the upper reaches, small chub (*Squalius cephalus*) hang in shoals of silver; in the lower reaches, sea lamprey (*Petromyzon marinus*) spawn. Chub don't have sea lamprey – also known as vampire fish – sucking their blood because sea lamprey only enter rivers with very clean water to spawn. Sea lampreys otherwise live and dine at sea, and chub stay in freshwater, so their habitat preferences keep them separate.

In 1957 the Thames was declared biologically dead. Its water was so low in oxygen that few species could survive in its anaerobic condition. Between its pure spring water start in Gloucestershire and emptying into the North Sea, the river was corrupted. A particularly sullen stretch of the Thames extended for a couple of miles either side of London Bridge, where the

< Chalk stream ripple with waterweeds and clear water

23

city and its people combined with damaged infrastructure were a problem. Victorian sewers that had been destroyed during Second World War bombings had not been replaced.

Standing on London Bridge now and looking down, the Thames water still looks murky brown. But it's a healthy shade of colour caused by particles stirred up in tidal sway. Rare but increasing sightings of seals in the Thames as it runs through the centre of London tell us that, unseen by most people, fish are swimming in the river and there are enough of them for a seal's appetite. The Zoological Society of London monitors fish in the Thames, and in 2016 found twenty-one smelt (*Osmerus eperlanus*) eggs at Wandsworth Bridge. The fact that these fish, notable for smelling like cucumber, were returning to spawn in the Thames was considered a marker of how much its water quality had improved.

Bringing the Thames back from the dead wasn't done just by sorting out London's sewers; wartime had put a strain on all our rivers and it took national effort to bring them back to life. During a session in the House of Lords in 1959, Pollution of Rivers and Estuaries was discussed. Fortunately, not everyone in the House of Lords or the House of Commons, who had previously discussed a bill on Clean Rivers (Estuaries and Tidal Waters), took the view of Viscount Simon, who inherited his peerage and power from his father. Viscount Simon stated, 'The natural channels for the disposal of waste in a country like ours are undoubtedly our rivers. People sometimes say (and I believe that this has been mentioned this afternoon) that this or that river is like an open drain. In my view, that is exactly what a river ought to be.' He continued to expound the view that rivers should be a clean and healthy drain, not a foul one, and in fact operate like man-made sewage disposal plants in terms of bacteria breaking down organic waste in the presence of oxygen. He bemoaned the overloading of rivers as natural sewage disposal systems, but did not think it necessary to purify sewage 100 per cent, as the river would deal with it. He commented, 'I do not want to go into technical details, and indeed I am incapable of doing so.' Fortunately for the majority of the population, who do not have hereditary power and have not scaled political heights, other members of the House of Lords were concerned about their fishing and did not agree with Viscount Simon. Lord Balfour of Inchyre, a keen fisherman, cited stakeholders including 2 million fishermen and said, 'Many of your Lordships have first-hand knowledge around the country of how fisheries and oyster beds are being poisoned'. He named other stakeholders, including residents and holidaymakers, who were entitled to better conditions, and reported that ninety per cent of cattle were drinking sewage-polluted water. He gave thanks to the kidneys of the cattle for their noble work in absorbing pollution and stated, 'sport, industry, amenities and health – the community are not getting the purity to which they are entitled'.

Various Lords gave examples of rivers in their holdings that were polluted, including an outbreak of typhoid because people had eaten shellfish from the Clyde estuary. In addition to sewage release, agricultural pollution via chemicals applied to crop fields, earth released by washing crops and 'synthetic detergents which are poured down sinks by so many housewives all over the country' were blamed for the state of our rivers. The Lords were well aware that river pollution was a recurrent issue, having previously been the topic of sixteen Acts of Parliament and fifty-four reports and investigations.

Clearly the House of Lords deplored the state of our rivers. The value of rivers was a matter of the economic cost of pollution and the fact that people were entitled to clean water. If fish were dead in the water, it was a sign that the water was not suitable for children to bathe in. Wildlife was absent from consideration, and so was any concept of a duty of custodianship and the idea of reciprocity – that natural resources are given to us but

we also carry responsibility for them.

Since 1857, when the Thames Conservancy, the first public body established to prevent river pollution, was created, successive sets of legislation have responded to pollution because it affects our lives. The Industrial Revolution and emergence of polluting industries on a vast scale led to the introduction of legal obligations for manufacturers to treat and dispose of waste water without contaminating rivers.

Innovation reduced some pollution by driving market forces in different directions. For example, the advent of digital photography caused a downturn in photo printing, and this reduced the release of silver particles used in photo development into freshwater. Industrial chicken factories have attracted attention for leaching excessive nitrogen into rivers, which causes algal blooms, reduces oxygen levels in water and kills wildlife. There are established means of dealing with manure that can be applied to deal with waste from chicken-rearing and egg-producing factories, and consumers can make choices on eating less chicken and selecting the conditions in which their chicken is reared. But as a whole, our society and its mechanisms of legal constraints remain orientated towards defending resources to ensure that we can use them. What happens if we change our view from entitled and

Trout (*Salmo trutta*)

transactional – on the basis of benefits we receive – to seeing wild waters as diverse inhabited places where we behave as guests who would like a return invitation?

Globally rare

In the patchwork of habitats coating the globe, a few rare examples are found in Britain and Ireland. Even in the relatively small space our countries occupy, despite the impact of population and industry, we are still custodians of special places.

Paradoxically, the cool running water of chalk streams is a result of the warm seas of 100 million years ago sustaining the life of creatures with calcium-rich bodies. During the Cretaceous period, the Earth's temperature and sea levels were high. Microscopic plankton called coccoliths inhabited the sea, and their skeletons rained down to the seabed. Those warm seas are gone, but their legacy of chalk shapes just over 200 streams and rivers in the world. Although chalk is present in several countries, only in England and

France is it close enough to the surface to form chalk streams. Most of the world's chalk streams are in England.

Streams arising from chalk and then flowing over sand and clay, as Hampshire's Itchen and Test rivers do, are chalk streams. When groundwater rises in greensand or clay and then runs over chalk, as the Stour in Kent does, it becomes a chalk stream. In East Anglia, Lincolnshire and Yorkshire, streams rise from chalk and run over gravel. In Dorset, Norfolk and the Yorkshire Wolds, streams run over chalk and then over greensand or Gault clay. In all of these rivers and streams we see the impact of chalk. Their water nearly always runs clear. Throughout the year they mostly flow close to their banks, but they have peak flow in spring. Plants thrive in their clear, mineral-rich water and are notable in green swathes in summer. Chalk streams tend to run shallow and relatively wide, rather than deep, for the volume of water they carry.

There is no need to put on goggles to see underwater in a chalk stream. Lie on a grassy bank or float on a paddleboard and without getting wet you can watch life in its water. If trout (*Salmo trutta*) didn't whizz by so fast, you could count the speckles on their flanks through water clear enough that it is almost as if there is no water. Ablaze with white flowers held just above the surface water, crowfoots do well in chalk streams and their slow-to-fluctuate water levels. Water-crowfoots don't grow well in murky water where they don't get enough sunlight, or in fluctuating water levels that swamp their flowers. Pond water-crowfoot (*Ranunculus aquatilis*) grows in winterbournes – streams that run dry in summer. Stream water-crowfoot (*Ranunculus penicillatus*) grows in chalk streams where water flow is fast. River water-crowfoot (*Ranunculus fluitans*) grows in lower sections of chalk streams where the stream or river is larger. Rare white-clawed crayfish (*Austropotamobius pallipes*) live in shallow water, and also need their water to be mineral-rich in order to fortify their hard exoskeleton with calcium. This makes chalk streams a refuge for these rare crayfish in England.

Gnarly pink seaweeds that have limestone skeletons form one of the world's habitats that supports great diversity of life. Coralline seaweeds – Corallinales – grow unattached to each other but their solid bodies interlock, creating sheltered space between them. Aggregations of these coralline seaweeds called rhodolith beds are, like coral reefs, a place for other wildlife to live. Rhodolith beds are a feature of marine landscapes from the tropics to the poles where water is shallow enough for light to penetrate and sustain life for coralline algae. Like peat, the fragile status of rhodolith beds is partly tied to their slow growth – damage to them can take longer than a human lifetime to heal. And both peat and rhodolith beds are vulnerable to climate change.

Rhodolith beds are called maerl around Scotland and England, and maërl in Ireland. In Scotland maerl beds are dominated by *Phymatolithon calcareum*, whereas in Ireland and England *Lithothamnion corallioides* can also be dominant. Growing only 0.2 millimetres per year, maerl around Scotland cannot shift to cooler areas faster than climate change is influencing sea temperatures. If emissions contributing to climate change continue at their current pace, by 2100 maerl beds around Scotland will decline by eighty-four per cent. Nearly one third of north-western Europe's maerl beds are in Scotland, so Scottish losses are also significant losses for Europe.

Maerl beds are an alien-looking world often within relatively shallow water, allowing people to see them while snorkelling. The top layer of a maerl bed is composed of living coralline algae, and underneath is old dead maerl or silty seabed.

Common brittle stars (*Ophiothrix fragilis*) – which look like skinny, hairy starfish – raise their arms in currents flowing over knobbly pink beds of maerl. Because maerl beds contain space between their units of coralline algae, oxygenated water can move through them, which allows wildlife that burrows to dig deep and remain within oxygenated space but be hidden from predators. Shellfish like queen scallops (*Aequipecten opercularis*) and truncate soft-shell clams (*Mya truncata*) hide their delicious flesh in maerl beds. Gravel sea cucumbers (*Neopentadactyla mixta*) conceal their soft bodies in maerl beds while keeping their feeding tentacles extended in the water to catch food.

On the Atlantic edges of Ireland and north and west Scotland, westerly winds blowing calcium-rich sand inland have created flower-peppered skeins of low-lying grassland called machair. Old seashells are responsible for the high calcium content of this sand. There are different varieties of machair, including cultivated and uncultivated, wet and dry. What unifies them is that their calcium-rich soil enables plants that need alkaline conditions to grow, and on much of the Hebrides this is a contrast to otherwise acidic soils. You can't be on machair without being close to the sea – it is a habitat on land but very much a product of marine life and oceanic winds.

Tenant farmers – called crofters in Scotland – created cultivated machair on which they maximised yields from their small-scale farms by rotating crops such as rye and oats with grazing by their livestock. Some patches of machair are maintained by occasional grazing but never ploughed for crops. Fertilised with animal manure and seaweed, these grasslands nurture small, slow-growing plants as well as grass. In summer the density of flowers on machair provides a rich source of nectar for bumblebees. A quiet visitor to machair can enjoy seeing waders that make use of it during nesting season – lapwings, redshank, snipe, dunlin, oystercatchers and ringed plover. Grazing and ploughing at low intensities are the other factors shaping machair; they maintain the characteristic balance of leafy plants and grasses.

Man-made habitats

There is no dispute that people destroy natural habitats, but sometimes they also create habitats that wildlife moves into. In the Anthropocene Age, wildlife that can make use of cities does well. A factor that anyone with a garden can influence is the availability of freshwater, by creating or maintaining a pond. Safe from agricultural run-off of pesticides, herbicides and fertilisers, garden

Garden pond in winter when grey wagtails visit and common frogs lay their eggs >

ponds can provide excellent water quality for a variety of wildlife. In winter, grey wagtails (*Motacilla cinerea*) will brighten your garden by flashing their bright yellow bellies while walking around the pond. These birds of hills and fast-flowing upland streams move to lowlands in winter. They eat midges, ants and tadpoles. For grey wagtails, a garden pond with a late winter crop of tadpoles is easy hunting.

Common frogs (*Rana temporaria*) spawn abundantly, but only a few out of hundreds of eggs will become mature frogs.

As well as being eaten at the egg stage, once hatched into tadpoles they remain an important food supply for birds and animals. On land, common newts (*Lissotriton vulgaris*) eat insects, slugs and worms; in water, they eat insects, water snails and tadpoles. Some dragonflies are exacting in their requirements and only wild water can provide the water depth, size of hunting territory, accompanying vegetation and in some cases acid water that they require. But many common dragonfly species like the broad-bodied chaser (*Libellula depressa*) and

common darter (*Sympetrum striolatum*) will breed in garden ponds. Dragonfly nymphs – immature dragonflies that live underwater for a year or more until they move to land ready for metamorphosis into adults with wings – fuel their growth by eating tadpoles. Soft-bodied tadpoles are rather defenceless and easy to eat once caught. Small tadpoles are even eaten by larger tadpoles. Frogs' profligate spawning supplies many creatures with food and manages to play the numbers game so that at least some eggs will reach maturity and go on to be adult frogs that spawn the next year.

In the aftermath of mineral resource extraction, valuable habitats can sometimes be created. In the last Ice Age, slow-moving glaciers eroded rock, reducing it to sand and gravel. When the climate warmed, melting ice turned into fast-flowing rivers that carried sand and gravel away, leaving it in concentrated deposits downstream where the river current had slowed. After gravel and sand, perennially in demand for cement and concrete manufacture, have been dug up via surface mining, large pits are left. These are often in river valleys – as they were deposited by old river courses – and where the water table lies above the depth of the pit, they naturally fill with water.

Many old gravel pits are now nature reserves and Sites of Special Scientific Interest (SSSIs). Being a water-filled hole in the ground provides a degree of protection by rendering land unsuitable for construction and agriculture. Gravel pits have unstable edges that create a mosaic of habitats. Deeper water in the centre with shallow edges caters to the needs of a range of animals and plants. The water that fills gravel pits comes from the water table. As such, it is filtered through the ground surrounding it. Gravel pits tend not to be situated in the middle of agricultural fields, which means that the freshwater entering gravel pits is a step away from the water running off fields containing nitrate and phosphate applied in fertilisers to maintain crop yields. Much as fertiliser promotes the rapid growth of crops, its residue in water prompts sudden algal blooms. Aquatic plants struggle to grow in water where light is blocked by algae. When algal blooms decompose, this reduces levels of oxygen dissolved in the water, making it impossible for aquatic invertebrates and fish to survive. In surveys of water quality, ponds and lakes have been less polluted than streams and rivers, with water in gravel pits particularly pollution-free. Gravel pits are a haven for wildlife that needs clean water.

Working quarries use pumps to prevent rainwater and groundwater that seeps in from accumulating. Once the quarry is spent, quarrying stops and pumps are removed. Ponds can develop on the quarry floor, and some quarries will fill to become deep ponds or lakes. Like gravel pits, quarries can be pollution-free oases of freshwater. Still water with a surface area over one hectare that has only moderate levels of nutrients is mostly confined to uplands in Britain and Ireland. Lowland quarries that are flooded artificially create similar-scale large bodies of water that are low in nutrients. Where limestone has been cut, rock cliffs and tunnels provide waterside accommodation for birds and bats. Quarries under different mineral influences than the surrounding landscape can offer refuge to otherwise unusual species. For example, in Derbyshire palmate newts (*Lissotriton helveticus*) are resident in Hadfields Quarry Nature Reserve, benefitting from the neutral water on the quarry floor, which is unusual in the limestone landscape surrounding it.

However, the subsequent conversion of a gravel pit or quarry into a nature reserve does not necessarily compensate for the damage caused by its creation. In the process of pit mining, wildlife inhabiting the space is killed or forced out by loss of its habitat. This isn't a Solomon's choice between grassland and forest, or gravel pit and quarry lake. It is an ongoing everyday decision to require resources buried in the ground for the construction industry, when their extraction has an impact on biodiversity.

Canals are green veins that pass through cities and otherwise sterile agricultural lands while brushing up against wilder places. Built as a transport corridor for goods, they have had a renaissance with canal-boat homes and use of their towpaths – wide enough for horses to walk on while towing boats on the canal – as car-free routes for walkers and cyclists through urban areas. For wildlife, canals aren't just a connecting corridor; they are places to live. Merits of canals include speed limits for boats, controlled and stable water levels, and slow water flow.

Stirred up by boat traffic and containing silt, canal water is rarely clear, but fish live there – enough fish to satisfy the appetites of otters (*Lutra lutra*). The recovery of otter populations has included the unexpected appearance of urban otters. Making use of canals for hunting and travelling, otters are regularly seen in Birmingham and Edinburgh. Swan mussels (*Anodonta cygnea*) burrow into mud at the bottom of canals. Usually found in standing freshwater of lakes and large ponds, swan mussels are able to live in canals because the water moves slowly. Scottish Canals leave eighty-seven per cent of vegetation on their land unmanaged. Canalside vegetation is a good place for grass snakes (*Natrix helvetica* subspecies *helvetica*) to hide and hunt, although you are most likely to spot these snakes while they are swimming in the canal.

Running water in rivers and streams

Perfoliate pondweed (*Potamogeton perfoliatus*) grows in both lakes and slow-moving rivers – much wildlife is shared between these two different types of water bodies. Yet there are differences. Perfoliate pondweed is at its most enjoyable in rivers, where it has the visual effect of a field of wheat rippled by wind, instead of rising in a stationary column to the surface, as it does in lakes and ponds. Moved by gentle river current, its stems sway without breaking, a leafy green movement that makes the current visible. Lowland rivers tend to be slower-flowing and are more often eutrophic (nutrient-rich) than upland rivers. Flowering rush (*Butomus umbellatus*) requires slow-moving or still water and can tolerate eutrophic water, so its upright columns topped with an umbel of pink flowers are a lowland delight. Kingfishers (*Alcedo atthis*) can be seen around lakes, and in winter will sometimes retreat to seashores, but lowland riverbanks are better able to cater to their need for exposed earth to dig nesting tunnels by water with surrounding vegetation of trees, shrubs and tall plants to perch on while watching for fish. Despite the velocity of most kingfisher sightings – a passing blur of turquoise and orange – they are better hunters in water that is slow-moving and doesn't have turbulence, which makes them more of a lowland river

bird. Roach (*Rutilus rutilus*) are classic lowland fish. Their habit of aggregating in shoals makes them easier to spot than solitary fish like pike (*Esox lucius*). Roach live in still or slow-moving water, more directed by a need for the depth and width of water characteristic of lowland rivers than for the well-oxygenated water that is more characteristic of upland rivers.

Upland rivers draining water from hills and mountains offer advantages and challenges. Their water tends to be pure and unpolluted, as little agricultural farming accompanied by fertilisers and herbicides, or industry releasing effluent – major polluters of lowland freshwater – takes place in uplands. Flowing down adds momentum and movement, which, in combination with the slightly cooler temperatures associated with higher elevations, keeps water well oxygenated. Wildlife that benefits from this high speed and clean water must also be equipped to deal with its challenges.

Upland streams and riverbeds are scoured by fast-flowing water, removing material like silt, sand and gravel. This means that plants need to be able to grow attached to rocks or tree roots. In upland streams and rivers, velvety green patches may be water-mosses. Alpine water-moss (*Fontinalis squamosa*) has pairs of leaf-like structures on stalks, looking rather like a trailing branch of weeping willow. It grows in fast-flowing acidic streams and rivers. Similar in appearance, willow moss (*Fontinalis antipyretica*) grows in slower-flowing and still water. Stoneflies – of which there are about thirty species in Britain and Ireland – rely on upland streams and rivers for their unpolluted, well-oxygenated water. Species such as predatory stonefly (*Diura bicaudata*) live in the water as larvae, shedding their skin in moults as they increase in size until they emerge as adults equipped for land, air and mating. Only spending a few weeks out of the water, and even then staying close to their stream or river, they spend their lives connected to upland water. Invertebrates like stoneflies are a food source for dippers (*Cinclus cinclus*), who are themselves adapted to life in fast rivers, with their ability to grip and walk on riverbeds while underwater.

Waterfalls offer some of the best salmon (*Salmo salar*) spotting, as leaping salmon reveal themselves airborne in their autumnal journey upstream to spawn. Waterfalls are also a feature that can make Britain and Ireland look tropical. On rocks around waterfalls, high humidity from splashing water creates a microclimate in which ferns and mosses thrive. Even the small-scale splash zone from a stream can make it possible for plants that need humidity to thrive. In 2020 a fern known

from Cuba, the Dominican Republic and Jamaica was found growing wild on rocks by a stream in County Kerry in Ireland. *Stenogrammitis myosuroides* is a Neotropical species that has managed to find a damp niche to grow in, thousands of miles from the climate it is accustomed to.

Standing water

Even amongst bodies of water that are still, there is not uniformity, which creates a range of conditions for wildlife in standing water. Oligotrophic lakes are low pH (slightly acidic) and contain low levels of nutrients. You can see the low nutrient level reflected in the clarity of their water, which is unclouded by microscopic growth fed by phosphate and nitrogen. Aquatic plants are sparse on their shores. Mesotrophic lake water can be clear, but the edges of these lakes can host many plants. Mesotrophic lakes are an unusual form, inhabiting a narrow balance of containing some nutrients, usually from natural sources, but not as many as eutrophic lakes. When lakes are rich in nutrients from natural sources or due to human activities, they are classed as eutrophic. Densely vegetated, and in summer often with green-tinged water due to algal bloom, they are lakes in which growth generated by nutrients is visible and rapid.

Water lobelia (*Lobelia dortmanna*) is absent from the south and east of both Britain and Ireland. It is slow-growing and cannot compete with other plants that grow fast when fuelled by nutrients in water, so it doesn't grow in eutrophic lakes. Rather, it is a plant of oligotrophic lakes and ponds. When the water is deep, it is often a companion to Arctic charr (*Salvelinus alpinus*), a fish that is a relic from the last Ice Age. In the northern range of Arctic charr in Scandinavia, they behave like their migratory relatives salmon and trout by moving between sea and freshwater. However, in Britain and Ireland they have stopped making this journey. Even in lakes where cohabiting salmon and trout suggest that migration back to the sea is physically possible, Arctic charr are not migrating to spawn. This isolation between Arctic charr in separate lakes has contained breeding pools to the extent that these lakes have Arctic charr populations that are genetically distinct. And these genetic differences are physically visible. They are not quite dissimilar enough to be considered different species by scientists today, although they have been classified as separate species in the past. If Ireland and Britain retain cold, deep lakes – despite the vulnerability of these lakes to climate change and nutrient pollution – it is possible that in time our Arctic charr may diverge

< Cool water with a lush tropical feel at waterfalls

enough for populations from different lakes to be considered different species.

Mesotrophic lakes are restricted to the edge of upland areas, and there are not many of them. With their characteristic balance of nutrients occupying a narrow range, they are vulnerable to change and are considered a threatened habitat. Lakes classed as mesotrophic can become eutrophic due to encroaching nutrients, and this alters the distribution of species. For example, slender naiad (*Najas flexilis*) disappeared from Esthwaite Water, one of the last sites it grew on in England, because of eutrophication. Slender naiad's branching form, which makes a green skein rising up from lake beds, still grows in some locations in Ireland and Scotland, and also in Northern Europe, but it is considered rare throughout its range. Loss from individual lakes can become death of a species by a thousand cuts.

Oligotrophic and mesotrophic lakes are home to some rare species of freshwater plants and animals. If you want to see reed beetles (*Donacia*) gleaming in metallic colours, oligotrophic and mesotrophic lakes are good places to find them. Their aquatic larvae eat submerged vegetation and also take oxygen from them. On leaving their juvenile life underwater, they live as mature beetles on reeds adjacent to the water. Mesotrophic lakes also host more common species. So within mesotrophic lakes there is combination of rare species with common species, which makes these lakes a concentration of aquatic biodiversity. Another element adding to the diversity hosted by mesotrophic lakes is their ability to provide conditions that accommodate species that otherwise do not coincide. Water milfoils (*Myriophyllum*) look like green feather dusters. While only committed botanists tend to bother to distinguish between water milfoils, the ability of a mesotrophic lake to sustain both acidic-loving alternate-flowered water milfoil (*M. alterniflorum*) and alkaline-loving spiked water milfoil (*M. spicatum*) is a biological equivalent of being ambidextrous.

Eutrophic lakes are not without charm, even if their water tends to be clouded by microscopic growth of algae in warmer months. A candidate for the biggest individual flowers you can see growing wild in Britain and Ireland, white waterlilies (*Nymphaea alba*), will make you feel as if you have stepped into a Monet painting. Their heart-shaped leaves rest on the water surface, interspersed with upturned floating flowers. They grow lustily, with their rhizomes (underground stems) absorbing nutrients from mud on lake or pond bottoms. Not just beautiful, they sustain life around them. Gently turning a leaf over, you will often see eggs sticking to the submerged side – dragonflies, damselflies and water

snails all make use of them. Their sturdy surface can be a convenient resting place for a frog, and a stepping pad for voles and moorhen chicks. Time spent quietly watching a patch of white water-lilies often gives the bonus of seeing wild animals as well as the pleasure of seeing those pristine and generously large flowers.

Ephemeral standing water is also an important habitat. Some rare plants survive on the edges of ponds that have fluctuating levels, or ponds that disappear entirely in summer. Pillwort (*Pilularia globulifera*) looks like grass, but is actually a tiny fern. It can survive aquatic and dry conditions, but cannot tolerate shading by other plants. While ground that is water-saturated and then dry is problematic for most aquatic or terrestrial plants, pillwort's ability to tolerate both states makes temporary ponds a place where it can thrive in the absence of competition. Natterjack toads (*Epidalea calamita*) spawn in shallow ponds where the water warms up in spring. Although these ponds tend to dry out and vanish in summer, they provide a home to natterjack toads during their aquatic phase of life as toadpoles. Ephemeral ponds contain fewer invertebrate predators – like dragonfly larvae – and they don't support fish, both of which eat toadpoles. Natterjack toads can take advantage of this only temporarily aquatic habitat because their young mature quickly.

Seaboard

Land under the influence of our sea takes a variety of forms. One of the most chimeric is salt marsh. Terrestrial plants and tides work together, forming a green carpet through which creeks cast silver threads. These creeks deposit silt, maintaining mud at a level where terrestrial plants that can survive occasionally being covered by salt water and growing in salty soil can take root. In eighteenth-century England, smugglers avoiding custom duty and tax on wine, spirits and tea found a niche in the salt marsh around Christchurch, Dorset. Smugglers who knew the marsh's creeks could quickly and discreetly bring goods across the harbour and into their collaborators' hands while avoiding customs officers.

Marsh samphire (*Salicornia*) is the first coloniser of seashore mud, and builds it into salt marsh. Its roots stop mud from being whisked away by moving water, and allow silty deposits to accumulate. Another early coloniser is cordgrass (*Sporobolus*, until recently named *Spartina* – botanists are still arguing about this name change that was made in 2014). Around Britain and Ireland, cordgrass is a mixture of *Sporobolus maritima*, which has been here since the last Ice Age, a nineteenth-century introduction from America called *Sporobolus alterniflora*, and their infertile offspring *Sporobolus* x *townsendii*. In an unusual toss of genetic dice, *S.* x *townsendii*

doubled its chromosomes and became fertile *S. anglica*, which is now the most widely distributed cordgrass in Britain and Ireland. Apart from being a reminder that the evolution of new species isn't confined to ancient history but is still happening, cordgrasses demonstrate characteristics that are well suited to salt-marsh living. They have two types of roots: deep roots that anchor them in the mud, and shallow roots that stay in aerated surface soil. Cordgrasses sweat salt through glands on their leaves, and you can often see salt crystals on them. Cordgrasses have thick-skinned leaves and their stomata – pores through which plants breathe – are in deep groves; both of these adaptations reduce water loss. They also deploy a water-saving mechanism mostly seen in plants that grow in the tropics. In cordgrasses, photosynthesis, the process by which plants convert sunlight into chemical energy, can take place without the stomata being fully open. By using this method, referred to as the C_4 pathway, cordgrasses release less water than via C_3, the pathway more commonly used by temperate plants, which requires fully open stomata and allows more water to escape. Much seashore wildlife has adaptations that enable it to survive the extreme and variable conditions which shape the environment along our coast, from high salt levels to temperature extremes and unstable ground.

Another grass has a role in maintaining those fantasy beaches of sandy shore backed by sand dunes. If you feel like you've been slashed by grass on a beach, it was probably marram grass (*Ammophila arenaria*). This grass becomes the skeleton of dunes, as its response to being buried by shifting sand is to grow. Marram grass can tolerate up to two metres of sand being deposited on it in one year. Growing up through sand deposition, marram grass creates dunes topped by spiky leaves with an interior in which old stems, rhizomes and roots stabilise sand. Sand dunes occupied by marram grass keep growing as they accumulate wind-blown sand and keep pace with sea-level rise. These dynamic dunes offer a natural means of moderating the impact of sea-level rise, in addition to their ability to reduce impacts of storm surges on land.

Behind the heights of stabilised dunes, hollows of low-lying ground called dune slacks provide shelter from sea breezes and offer the combination of wet in winter and dry in summer conditions that some of our rare plants and animals thrive in. Many populations of natterjack toads live in temporary ponds that form in dune slacks. Although marsh helleborines (*Epipactis palustris*) have declined in their range, dune slacks are one of the places they grow, by making use of seasonal water and drought

that deters competition. Spreading by rhizomatous roots, they can form patches so dense that it takes a moment to realise the ground is solid with frilled white and red petals. Critically endangered fen orchids (*Liparis loeselii*), having lost much of their eponymous fenland habitat of low-lying marshy ground to drainage and agricultural use, have a refuge in damp dune slacks. At Kenfig in Wales, dune restoration has brought back these pale-chartreuse flowered and diminutive – eight-centimetre tall – orchids in their thousands.

The height of a clifftop does not offer insulation from maritime influence; wildlife adjusts to high winds, salt spray and low availability of freshwater here too. Buck's horn plantain (*Plantago coronopus*) can tolerate salt and grows wild on rocks, dry grassland and clifftops. On cliffs by seashores, what looks like short grass is often on closer inspection – achieved by undignified crouching, kneeling or crawling to peer at the ground – a blend of small and finely leaved plants including buck's horn plantain. Dwarfed by tough conditions and being cropped by grazing livestock, its leaves are thick enough to have succulence. Not only delicious to animals, it is also a wild plant used in salads in Italy, and has become a garden-grown vegetable called star grass, *erba stella*, or minutina – although in gardens provided with richer soils and regular watering, it grows much bigger than in its wild clifftop life.

Short grassland on sea cliffs is essential for choughs (*Pyrrhocorax pyrrhocorax*), the bird on Cornwall's coat of arms. In 1973 the last Cornish chough died, and although there were occasional sightings, no breeding pairs were observed in Cornwall. Living on insects and nesting on rocks, these rare birds – there are only about 1,400 breeding pairs in Britain and Ireland – do well on land that is used for low-intensity grazing. They find insects in and on the soil, which is easier to peck at when grass is short. They can find food in winter dung left by grazing animals and in seaweed along strandlines, even if the ground is frozen. Although in the seventeenth century choughs were thought to be common, their populations subsequently declined due to trophy hunting, and also because when livestock was moved inland, scrub and taller plants encroached on clifftop vegetation. This made it harder for choughs to find enough insects to eat. Work on restoring populations of chough uses livestock grazing to bring back the habitat that choughs need, which also recreates a maritime grassland habitat that is important to other species of wildlife such as wild asparagus (*Asparagus prostratus*), the silver-studded blue butterfly (*Plebejus argus*) and the greater horseshoe bat

(*Rhinolophus ferrumequinum*). The return of choughs to Cornwall was happenstance. Three birds arrived in 2001, two of which paired and successfully reared chicks in 2002. Their numbers in Cornwall have been increasing slowly. Within the natural range of animal pairings, which is not restricted to heterosexual partnerships, in 2007 two male choughs paired and for several years built a nest and held territory together. Cornwall has embraced the return of its bird that holds the spirit of rocky shores lapped by gently grazed land, and chough-friendly habitat management is being encouraged.

Sheer walls of sea cliffs provide birds like choughs with space to nest in that is safe from terrestrial predators such as rats and house cats, although they are vulnerable to predatory birds. They are also a place where plants equipped to grow in tiny scraps of soil caught in crevices are found. Golden samphire (*Inula crithmoides*) is tolerant of salt and grows on salt marshes and cliff faces. Its succulent leaves, which are filled with water, buffer it from water shortage. You are more likely to notice it when you see its yellow flowers and yellow–green leaves on sea cliffs and salt marshes, but if you read the labels of skincare products you will sometimes find it listed as an ingredient. Golden samphire is promoted as a regenerating and protective agent in cosmetics.

Pay attention to rocks from the top of a cliff to its base and you may be surprised at the diversity of lichens that can coat them, sometimes to the extent that what appears to be bare rock is entirely covered by lichen. Sea ivory (*Ramalina siliquosa*) stands out in pale grey–green tufts. It often grows in juxtaposition with maritime sunburst lichen (*Xanthoria parietina*), which lies flat against rock in bright orange patches. Lichens even reach into the sea, with black lichen (*Lichina pygmaea*) in scruffy mats at the upper edge of patches

of barnacles where high neap tides cover them both with the sea.

Shingle shores can look dull from a distance, but close up there is life amongst the loose stones. Yellow horned-poppy (*Glaucium flavum*) is not just yellow in flower – if you break a leaf or stalk, lurid yellow sap wells up. It overwinters as a cluster of small hairy leaves, and in summer its leaves burgeon into waxy leaves that are less hairy. Hair reduces water loss by reducing the desiccating effect of wind, and waxy leaves are better

at keeping water in. Despite the abundance of water on seashores, lack of freshwater is a limiting factor for terrestrial plants, so in order to be successful they must be efficient at retaining freshwater.

Shingle beaches have a scattering of rare insects. Whelk jumping spider (*Pellenes tripunctatus*) is named after the whelk shells that this jumping spider hides its egg sacs in. Empty whelk shells washed up by the sea are often found on shorelines, and whelk jumping spiders take advantage of the shelter they provide. Ringed plover (*Charadrius hiaticula*) are ground-nesting birds that lay their clutches of eggs in stony nests. If too much vegetation grows on a site they abandon it. This is because their eggs are patterned to be camouflaged on a stony background; surrounded by vegetation, their eggs are more visible. Shingle beaches provide a stony background for their nests in close proximity to the marine worms, crustaceans and molluscs that they feed on.

Rock pools offer safety from big predators of the sea, but there is a price: extreme swings in temperature as small bodies of water heat in a day's sun and chill overnight. Immersion in water is accompanied by occasional exposure – it is a challenge to survive both states. Perhaps rock pools are also the most rewarding

place to see wildlife, as they offer a theatre including easy-to-spot immobile creatures as well as fast-moving wildlife that is trapped within your field of view.

The first rule of rock pools is that if you lift stones to look at what is on or under them, replace them exactly as you find them. The second rule of rock pools is the less disturbance you create, the more you may see as their inhabitants forget you are there and continue to go about their business.

Anemones with their soft bodies attached to rock are animals that look like flowers, though these flowers are visible year round. Strawberry anemone (*Actinia fragacea*) resemble strawberries when their yellow-speckled bodies are marooned above the low tide. At high tide when the sea returns, their tentacles emerge and they look like spiky red flowers. Dahlia anemones (*Urticina felina*) rampage through different colour combinations from one individual to the next, but their chubby tentacles very much look like dahlia petals.

A real fish of rock pools, more than a marine fish that happens to be in a rock pool, the shanny (*Lipophrys pholis*) can even hide above the waterline under seaweed or damp rocks at low tide, waiting for high tide to return. Individuals vary immensely in markings, from pale with dark speckles to near-solid colour in dull shades of grey and brown.

< Beadlet anemones and edible crab amongst wrack and rockweed in a rock pool

However, their colouring tends to match their surroundings, so when they are motionless they are hard to spot.

Marine life

Rocky shores segue into rocky reefs, and the blend of seaweeds and animals that live on them changes with the transition from rocks that are briefly submerged by tides to rocks that remain underwater even at low tide. Gullies and canyons concentrate the force of water moving under tidal influence and are ideal places for filter-feeders like plumose anemones (*Metridium dianthus*) with a firm grip to keep their place and catch passing food. Chalk and limestone reefs are soft enough for wrinkled rock-borer (*Hiatella arctica*) to dig itself into them. Where water is clouded with particles and light is limited, rocks are covered by animals like molluscs, sponges and anemones; in clear water lit by sunlight, seaweeds take hold.

Too big for the average rock pool, kelp unfurls in carpets called kelp beds on the sea floor. These forests of kelp underwater take up as much space as our broadleaf forests on land. Trees have roots, trunks and leaves; kelps have holdfasts, stipes and blades. Holdfasts are tentacle-like domes that grip on to rocks and resist

< Layers of life in a kelp forest

being torn away by currents. Brittle stars and worms can be found living inside the shelter of holdfasts. Cuvie (*Laminaria hyperborea*) skirts Britain and Ireland along their rocky shores. Its stipe has a rough surface that makes it easy for red seaweeds like dulse (*Palmaria palmata*) to attach and grow on it. Animals such as sea squirts, bryozoans, hydroids and sponges also settle on the stipes. Individual cuvie kelp can live up to fifteen years, providing relatively stable accommodation for the life that attaches to it. Oarweed (*Laminaria digitata*) also forms kelp beds, although its smooth stems remain mostly uncolonised, with just older stipes that may have roughened with age having dulse growing on them. Warming sea temperatures are ushering in golden kelp (*Laminaria ochroleuca*), a seaweed which previously had a Mediterranean range. Golden kelp, like oarweed, seems to be bare of the life that attaches to cuvie kelp.

Seagrass beds have attracted international attention for their previously overlooked ability to sequester carbon. Looking very much like grass, the two dominant types of seagrass around Britain and Ireland are eelgrass (*Zostera marina*) and dwarf eelgrass (*Zostera noltii*). Both species have slim green leaves between which small creatures can rest and forage whilst being concealed from larger predators. For short-snouted

Short-snouted seahorse (*Hippocampus hippocampus*)

(*Hippocampus hippocampus*) and spiny (*Hippocampus guttulatus*) seahorses, a leaf of eelgrass is a perfect anchor to wrap their tails around without causing any damage. Conversely, seagrass beds are damaged by boat anchors and this has a negative impact on our seahorse populations. Despite being feeble swimmers, greater pipefish (*Syngnathus acus*) – similar in appearance to seahorses except their bodies are straight – can manage to move around seagrass beds as they are a place where water flow is slowed. Seagrass beds are well stocked with crustaceans and soft-bodied invertebrates that resident fish can eat. A particularly ornamental group is stalked jellyfish (Stauromedusae), each one a

tiny composition of curlicues. Seagrass beds are also important refuges for bigger predators when they are juvenile and the prey of other species. Blades of eelgrass offer a useful anchor point for shark egg cases and cuttlefish eggs.

Seagrass and kelp beds and rocky reefs are a treasure trove of life that spills out into the open ocean – because of the shelter they offer juvenile fish. Damage caused to these places by boat anchors, mechanical harvesting of kelp and dredging for molluscs has an impact on life offshore, including some that appears on our supermarket shelves and plates. We have not looked after our seagrass meadows. Since the 1980s we have lost forty-four per cent of seagrass coverage around the United Kingdom. Looking back over a longer time span and potential distributions of seagrass, we may have lost ninety-two per cent of these marine meadows, and the accompanying loss of wildlife that inhabits it.

Pollack (*Pollachius pollachius*) live in shallow water among rocks and kelp beds, feeding on crustaceans when they are young. When they reach about three years old, they move out to deeper waters to feed on open-water fish and deep-sea prawns. Atlantic herrings (*Clupea harengus*) spawn in coastal areas but live their lives offshore in shoals. They can swim down to 200 metres deep, but more spectacular is when they shoal near the surface and they cannot be counted by the thousand, but rather measured by how many square kilometres their shoal covers.

Cetaceans, fish-shaped in body but in fact mammals at home in the sea, roam our coasts and surrounding seas. From land you can spot both relatively small common dolphins (*Delphinus delphis*), which are about two metres long, and orca (*Orcinus orca*) up to nine metres long. Travelling between our islands by ship, you can see long-finned pilot whales (*Globicephala melas*) – a small whale being perhaps only four to six metres long – and even fin whales (*Balaenoptera physalus*), which at up to twenty-four metres are significantly longer than a double-decker bus.

Aquatic life has generalists who appear in many places, and specialists who are specifically adapted to exploit niches where competition is small because of the environmental challenges limiting the field of contenders. A clear-running chalk stream is not just a place to see speckled trout whizz past and water crowfoot cover the surface with flowers. A chalk stream is more beautiful because of the wildlife we see in it. As much as an ocean lover's pulse quickens at the scent of salt in the air, we can be triggered by the visual transition of lichens in grades of salt and spray tolerance that leads from terrestrial to submerged. In a kaleidoscopic shuffle of elements when we look into a river or rock pool, alongside the common and expected features there is often a local quirk of geology or custodianship that means less frequently seen wildlife also has a home.

3

The Time Dimension

Somewhere revisited at different times of day or year becomes a foreign country in which new sights are revealed. It takes six hours and twelve minutes for shoreline water to rise from low to high tide. Mayflies dance out their adult life in one day. Because wild life and wild water are not static, our own doorsteps can be exotic if we break our routine expectation and go there at a different time.

Temporary waterways

Winter is not just the time when ephemeral ponds are topped up by high rainfall and low evaporation. It is also a season that fuels temporary streams. Looking at a map

< Enjoying bioluminescence in summer seas

of southern England, you can see places called 'winterbourne' – Winterbourne Steepleton in Dorset and Winterbourne in Kent. They carry the name for a feature common on chalk: winterbourne streams that run dry in summer. Even in the dry phase, the water channel is visible, slightly indented in the ground. Patches of gravel can be seen, and in some hollows a puddle of water might remain, containing damp mud with vegetation growing on it that remains a little greener than the surrounding grassland. As autumn turns into winter, little leaves appear between blades of grass. These belong to seedlings of watercress. Invisible to the surface view, water is increasing in the stream channel's substrate as the groundwater that feeds the winterbourne is recharged by rain. On a wintery visit, puddles that were previously a thin layer of water on mud start to have depth and stretch along the stream channel. A trickle of water starts to connect puddles, and aquatic plants burgeon into growth, watercress growing over grass.

Some fish move into winterbournes as they flow. Brown trout (*Salmo trutta*) take advantage of the gravel beds that winter-fed winterbournes offer, free of organic matter and silt. They use gravel beds as a place to spawn, and leave their eggs in an environment in which hatched juveniles face less competition from other fish for invertebrate food. As water flow declines, juvenile trout travel downstream to permanent bodies of water. Bullheads (*Cottus gobio*) are opportunistic and move upstream when water is flowing in winterbournes. They can become trapped in pools as flow drops and offer a feeding bonanza to grey herons (*Ardea cinerea*) and other fish-eating birds. Those not picked off by predators die as still water decreases in oxygen content, and eventually pools dry out. Winterbournes host otherwise rare insects such as blackfly (*Metacnephia amphora*), stonefly (*Nemoura lacustris*) and soldier fly (*Oxycera terminata*). Because of the number and rarity of species of aquatic invertebrates found in winterbournes, they are considered to have high conservation value. However, chalk streams that turn into winterbournes due to excessive extraction of water from underlying

aquifers are not a conservation gain, rather a sign that water use is unsustainable.

Winterbournes are a feature of our lowland chalks; mountains have their own seasonal flow. Unlike winterbournes, which are fed by groundwater and have predictable summer drought, headwater streams in upland areas are determined by rainfall and snowmelt. These streams of steep slopes are more characteristic of mountainous and hilly regions of Wales, Scotland, northern England and Ireland.

Brown trout will travel upstream when water is flowing in headwater streams, but few other fish have the agility and speed to progress against the terrain and the rapidity with which spate can flow. Even brown trout are occasionally caught out and can be stranded in pools as the flow drops.

Another type of temporarily flowing water is formed in karst landscapes where rivers run across bedrock that is soluble. The passage of water and time creates underground drainage. In peak flow, these rivers run despite drainage; but in times of drought, lower water levels prevent them from continuing to flow. These karst landscapes are seen in the west of Ireland and in England's Peak District and Yorkshire Dales.

In dry phases, karst riverbeds are inhabited by ground beetles such as green and polka-dotted *Elaphrus riparius*. This beetle prefers habitat that includes sparse vegetation, although it requires soil that is permanently damp to live in. Dry karst riverbeds meet their needs by providing both open areas and soil that remains damp. Rotting algae and debris left from when the river was flowing are rich places for these beetles to scavenge for food. Unexpectedly, some aquatic species can survive without water for months. Pea mussels (*Pisidium tenuilineatum*) can survive in sediment left on riverbeds that are dry for up to eight months. While life persists in these temporary rivers during dry phases, the longer the duration of drought, the fewer species survive it.

Toads (*Bufo bufo*) and great crested newts (*Triturus cristatus*) spawn in ponds fed by temporary flow and pools that form as they dry out. The density of toadpoles in temporary pools can provide a bountiful feeding opportunity for grass snakes (*Natrix helvetica*), which move fluidly between terrestrial and aquatic hunting grounds. The benefit for toadpoles in these pools is that they face less predation by fully aquatic species than they would if spawned in permanent ponds. Whether wet or dry, stream and river channels are used as corridors in bat migrations. Temporarily dry riverbeds offer a patch of rich plant growth inhabited by insects, ideal foraging for both herbivores and carnivores like shrews (*Sorex araneus*).

Drying out temporarily isn't

necessarily a bad state for a stream or river. It does have an impact on the wildlife that inhabits the space, and can allow some species a niche in which they can thrive while their competition is reduced. Wildlife in temporary rivers has not received equal attention in all the places where these rivers occur. Winterbournes of the lowlands have been studied more than karst rivers and upland headwaters. This is an artefact of academic and logistic bias, yet also an opportunity for everyday people who explore on their doorstep to be at the forefront of ecological knowledge.

Seaweeds in winter

On land, spring tentatively puts out a few leaves, but it is not until May that deciduous trees are fully clad. On grey winter days, the place for lush fresh growth is in the sea. People are more active in the sea in summer, particularly from June to August, when the water is at its warmest. This might be the reason why our seaweeds are underappreciated, as many are at their best in cooler months.

Cuvie kelp (*Laminaria hyperborea*) starts to grow its new blades in November. Over the course of months being tossed in sea currents and nibbled on by herbivores, old blades of cuvie become frayed at the edges and overall look quite worn. In spring and summer, just before the old blades are shed, cuvie kelp beds look tattered and dull. They are in their prime attire in autumn, when old blades are shed and the rippling fronds on display are all looking fresh.

Dabberlocks (*Alaria esculenta*) has its fastest growth rates in spring. By June its growth slows. Wave action can tear its blade back to the central midrib, leaving it looking shredded. High water temperatures and the stronger bleaching action of sunlight in summer erode patches of dabberlocks that are in the intertidal zone, where they are easiest for us to see. It is another seaweed better appreciated in cool months.

Sea beech (*Delesseria sanguinea*) is a crimson seaweed that looks like elongated leaves of copper beech. For sea beech, new growth starts in February and blades reach their full size around May and June. Over the rest of the calendar year, the blades accumulate tears and marks to the extent that by December they can be reduced to just midribs. For many seaweeds with their cycle of discarding old fronds, it is in winter and spring when we see them in their peak of fresh growth.

Deciduous phases

While late winter and early spring burgeons in the sea, on land there is a scattering of seasonal highlights from waterside plants. Many river walks in the cold months of November to January are paused due to a strong, sweet scent in the air, whose origin is not obvious. Winter heliotrope (*Petasites fragrans*), which is often found growing on damp ground along riverbanks, has inconspicuous but strongly scented flowers.

In summer, alder trees (*Alnus glutinosa*) blend into the blur of green leaves that surrounds us. In winter, alder stand out, marking the damp ground they grow in with a purple haze cast around their bare twigs. They are covered with dusky purple catkins held on pink-flushed stalks. Alder are sensitive to drought, so their distribution hugs wet places. They mark out edges of rivers, lakes, canals and ditches, and grow in dune slacks, flood plains, fens and bogs. They also create carr – wet woodland – while growing en masse in waterlogged soil.

On a shingle seashore, there is often an unexpected dash of purple, but here you have to be close to see it. Between stones, the crumpled leaves and stalks of sea kale (*Crambe maritima*) emerge purple. They will burgeon into large grey–green leaves topped by a cloud of honey-scented flowers in summer.

Coltsfoot (*Tussilago farfara*) puts up bright yellow flowers before its leaves emerge. Its aquatic location is not because of a need for moisture – it is drought-tolerant – but it grows by rivers and on seashores because they share the feature of disturbed ground. Crumbling riverbanks, slumping soil at the base of cliffs, landslides, shifting sand dunes and shingle all offer the chance to grow with little competition. Coltsfoot seeds, like those of its close relative dandelion, are each equipped with a parachute of fine hairs that catch even light breezes. This lets coltsfoot travel to new places. Spreading vegetatively via rhizomes allows it to keep purchase in unstable land. While most terrestrial plants are still dormant, coltsfoot's bright flowers stand out against bare ground as early as February.

Looking at maps of the distribution of alexanders (*Smyrnium olusatrum*), you can see its preference for coastal areas of Britain and Ireland, and a southern skew with very little occurrence in Scotland and northern England. By seashores this Mediterranean introduction – brought here by the Romans – is buffered from freezing weather by the moderating effect of the sea. Where alexanders grows inland, it tends to be associated with sites of old human habitation, having been deliberately nurtured for its use as a vegetable. Intolerant of prolonged frost, alexanders leafs up in the cold months. Clumps of shiny green leaves in January to March stand out against plants waiting for warmer weather. Leafing up in winter is a habit many Mediterranean species use

to take advantage of cooler months when water availability is better, leaving summer months as a time of maturing rather than a phase of active growth.

In summer, large pink flowers highlight the presence of a two-metre-high plant that lurks along many watercourses. Its height is even more remarkable for representing growth from small seed to person-sized in one season. Himalayan balsam (*Impatiens glandulifera*) is our tallest annual plant. It is most commonly noticed on the banks of waterways, but it grows in damp woodlands, mires and flushes – where upwelling water creates saturated ground but not a channelled stream of water. With bright pink, large and unusually shaped flowers – which have inspired its other names of policeman's helmet and gnome's hatstand – it is unmissable. Originating in the Himalaya, it was introduced to Britain in 1839 as a garden plant, and by 1855 was growing wild in Middlesex. After escaping from gardens in other locations, it then spread with assistance from our waterways. For Himalayan balsam, streams and rivers are transport routes, as its seeds are carried by water flow. Where they are washed into soil, they grow and establish in that new location the next year. As their seeds are slingshotted out of their capsule when ripe, these plants don't have to be directly next to flowing water for the seeds to gain access to streams and rivers as transport for future generations. In the Himalaya, it grows mixed amongst other plants; in Britain and Ireland, it forms dense stands. Few of our local plants can compete with its speed of growth, so it is considered a problematic naturalisation, but certainly a plant that anyone enjoying our riversides will meet. Himalayan balsam is a plant that marks the seasons with its rapid summer growth, as well as being notable for its quick transition from garden plant to being established in the wild.

Another plant that surfaces to attention in summer is water pineapple (*Stratiotes aloides*). From autumn through to spring, a glance in a lake, pond or ditch might include seeing what look like submerged pineapple tops. Like its namesake, the leaves of water pineapple are serrated and hard. In summer, water pineapple rises up enough for its leaf tips to poke out of the water and for its flowers to emerge in air. As a dioecious plant, it needs insect pollinators to bring pollen from male plants to the flowers of female plants in order to set seed. This is not possible in Britain or Ireland, where all of the plants are female. Instead, water pineapple here spreads through vegetative reproduction, with small clones snapping off the main plant. It was first recorded here in 1633 and, after increasing to spread in calcareous and meso-eutrophic bodies of water, has declined in the past 150 years, probably due to an increase in phosphate levels from sewage, slurry and detergents.

For most plants, autumn is a time of drawing down into seeds and roots. Some plants remain green, insulated from cold winter air by their covering of water. Several seashore plants, such as sea beet (*Beta vulgaris*) and rock samphire (*Crithmum maritimum*), are evergreen, and several hardy annuals like Danish scurvy grass (*Cochlearia danica*) get growing over winter as they benefit from the seashore's tendency to remain frost-free. After being marked by a blaze of autumnal yellows and oranges, forests shedding their leaves reveal new views. Animals and birds that don't migrate or hibernate for winter are much easier to spot without green leaves to remain concealed behind.

Seasonal animal behaviour

In September, swallows (*Hirundo rustica*) and house martins (*Delichon urbica*) gather in groups, often perching arrayed on phone wires. They migrate in flocks to overwinter in Africa. We know summer is returning as on warm days in spring, swallows and house martins become a frequent sight around lakes and rivers. Throughout summer these birds take advantage of the high numbers of insects that hover around freshwater. They swoop and hunt, catching insects as they fly. In late spring these places have an added draw as a source of mud to collect and use to repair old nests or build entirely new ones. It takes around 1,000 beakfuls of mud to make one nest. Their close relatives sand martins (*Riparia riparia*) don't build nests; they burrow into high riverbanks and sea cliffs made of sand or earth. Sand martins also feed on the diversity of insects that gather in air around wetland habitats. Swallows, house martins and sand martins sip freshwater on the wing as they skim over the surface of lakes and rivers. As they spend their summer in constant motion or hidden in nests, their autumn gatherings before migrating are an uncommon opportunity to see them stationary.

As the year begins, rising tides carry glass eels (*Anguilla anguilla*) into estuaries. At this early stage of their lives, European eels are only six to seven centimetres long, and look like transparent worms with eyes. You can see their internal organs through their invisible flesh. It is easier to sample glass eels entering rivers in their juvenile migration than to count mature eels on their journey to the Sargasso Sea where they spawn. The number of glass eels reaching our estuaries from their origin in the Sargasso Sea has decreased dramatically, and this is thought to be due to a decline in the adult population. European eels are critically endangered. Their appearance changes over the

course of their life. As they begin living in freshwater, their bodies become opaque with pigment and they are known as elvers. Once over seven centimetres in length, they are called yellow eels – as although their upper side is dark, their bellies are yellow – and this is the form they have for up to twenty years as they live and mature in rivers and lakes. Heading back into the sea, their bodies become silvery and they are called silver eels. It is possible to see mature eels hiding amongst waterweeds in rivers by daylight. Glass eels are easier to see, as they form active wriggling crowds heading through estuaries and into the lower reaches of rivers in January, February and March at night. Taking advantage of spring tides, they are swept upstream; as the tide ebbs, they mass close to riverbanks, within view of torchlight from riverside footpaths.

Summer's end brings silver clouds of fish close to shore in the form of shoals of sprat (*Sprattus sprattus*). For much of the year, they lurk around coastal waters in small shoals as they feed on plankton. As sea temperatures start to fall and day length shortens, sprat gather in larger shoals. Unfortunately for sprat, the size of their shoals makes them detectable by sonar. Sprat fishing season runs from August to February. Sprat migrate between inshore winter feeding and offshore spawning grounds. Around Britain and Ireland, peak spawning takes place in summer from April to August, although in the Mediterranean, peak spawning is in the winter months. In autumn it is not uncommon to encounter sprat shoals on a sea swim. Or, if they have been chased by predatory fish and taken evasive action in panic, they can sometimes beach themselves in large numbers, leaving hundreds and even thousands of small fish dotted above the waterline.

Brown trout (*Salmo trutta*) return to their natal patch of river to spawn – i.e. they spawn in the place where they were spawned. They are joined by sea trout (also *Salmo trutta*) which were spawned in the same place but after hatching moved into the sea. Brown trout exclusively live in freshwater. Spawning is the time when these different groups of the same species meet and interbreed. The more numerous brown trout have their next generation invigorated by the contribution from the less numerous but fitter sea trout. Sea trout produce more eggs, and they have larger yolks, which means that they hatch bigger and sooner. It isn't clear what causes a brown trout to either remain in freshwater or move to sea and become a sea trout. Brown trout and sea trout remain connected by their place of spawning. Between November and February they seek out gravel beds in rivers and streams to spawn. Occasionally they can spawn in lakes where groundwater provides the flow

of well-oxygenated water that their eggs require to develop. Trout spawning time is a moment to spot trout in groups from riverbanks. It is also a reminder that over the following couple of months, gravel beds in rivers should be left undisturbed to allow trout eggs to hatch successfully. Similarly, their close relative salmon (*Salmo salar*) makes an autumn migration for spawning in gravel on riverbeds. They can be seen in the river, and also in full view as they leap out of the water to travel upstream through waterfalls and past obstructions. Increasingly, dams are built with salmon ladders, a series of pools that help them to scale the height of these man-made objects between them and their spawning ground.

While fish spawning times vary, and in some cases can occur throughout the year, birds predominantly nest in spring. A few species produce second and even third broods in summer, with some urban birds, such as pigeons (*Columba livia domestica*), even occasionally nesting in winter.

Fulmars (*Fulmarus glacialis*) live offshore but come inshore in May to nest on sea cliffs. As the tactical advantage of being in a cliff limits predators that might harm their eggs and chicks to those that fly or can scale cliffs, fulmars have an additional defence. A climber arriving on

Fulmars nesting on sea cliffs >

a fulmar nesting ledge will be unwelcomed with a vomit of stinking, partially digested fish hurled at them. Other nesting birds are not equipped to defend their nests in this way. They rely on making alarm calls, circling in the air and dive-bombing intruders. Cliff-nesting birds that live in large colonies can mount loud and effective defences as a group. Seeing these behaviours indicates that birds have been disturbed, and it may be that you are too close and causing birds distress. Nesting is better watched through binoculars, spotting scopes or nest cams, all of which allow birds to remain undisturbed by people enjoying the spectacle.

Ground-nesting birds such as redshank (*Tringa totanus*), which nest on salt marsh, are more vulnerable than cliff-nesting birds. Most people will easily avoid accidentally stepping on a nest. But from spring until chicks have fledged in summer, canine companions are best kept on a lead, as given the option of an egg or chicks they might take it. Free-ranging dogs, even if they don't directly harm eggs and chicks, cause parent birds anxiety and disruption as they watch interlopers and try to distract them from their precious nest. Eggs that need incubating can quickly cool if parents have to leave the nest to deal with perceived threats. Chicks that need feeding can lose out when parents are busy defending their nest instead of collecting food for them. Many nature reserves will post signs alerting people to the presence of ground-nesting birds and the need to keep dogs on leads. But even in the absence of signs, we can watch and listen out for birds deploying distraction techniques and keep ourselves and our dogs on paths.

Nocturnal behaviour

Exploring outdoors by day is our default, so we miss animals that orientate their behaviour towards darkness rather than sunlight. Navigating outside at night is aided by having been there in the daytime so that you are familiar with the terrain, your intended route and where you might change your route. Certainly, urban lighting can help you to see wildlife around canals and ponds, but the taint of electric light spilling from houses and street lights also steals our view. In a town centre we might see 100 stars; under a dark sky 1,000. Not all night skies are equal. Heading away from built-up areas leads you towards darker skies.

Turning away from towns glowing in the distance and looking towards dark horizons helps. Looking out across open sea is another step towards darkness. Around Britain and Ireland, there are several places that have been internationally recognised as Dark Sky Reserves. Here, starry nights and nocturnal life are protected. Ecosystems are affected by light pollution. Birds like woodcock (*Scolopax rusticola*) that migrate at night can collide with windows, thinking reflections show open sky, not solid building. Nocturnal seabirds including shearwaters, petrels, and albatrosses (Procellariiformes) are disorientated by bright artificial light. Land managers in Dark Sky Reserves implement long-term planning strategies and regulations that limit light pollution in order to maintain a dark sky view and to control light in the peripheral area surrounding the dark sky. Using a torch, especially if you are careful to avoid dazzling animals by shining light directly at their eyes, is not a big disruption to animals. However, human eyes adjust to low light levels. Even waiting in the dark for ten minutes will make a difference to how much you can see without needing a torch. On clear full-moon nights you will even find that the moonlight is bright enough to make you cast a shadow. It is worth being equipped with a torch, but try giving your eyes a chance to get accustomed to night-time views.

In summer's peak of July and August, an evening waterside walk with a torch may draw moths. Beautiful China-mark moth (*Nymphula stagnata*) isn't just on the wing as an adult around lakes, ponds and rivers; it is an aquatic moth. Its larvae live underwater, where they hibernate inside the stems of their food plants – yellow water-lily (*Nuphar lutea*) and burr-reed (*Sparganium erectum*) – and then eat them from within the stem. Ringed China-mark (*Parapoynx stratiotata*) larvae are more active in water, enabled by their branched gills that are more efficient at absorbing oxygen. These moth larvae feed on smaller plants without stems that they fit inside, such as hornworts (*Ceratophyllum*). Most female water veneer moths (*Acentria ephemerella*) are so aquatic that they don't fully develop wings as adults. Instead, they live on the surface of water or just below it. Fertilised at the surface by male water veneer moths – who do live on the wing as adults – a female swims down to lay her eggs on aquatic plants. Their larvae chew into plant stems and build shelter by gluing pieces of plants together. Water veneer's preferred larval food plant is water milfoil (*Myriophyllum spicatum*), which has led to the suggestion that water veneer can be used as a biological control for this plant in North America where it has been introduced. However, like all biological controls, there is a risk that the intended plant will not be the only target.

Many of the charismatic animals we think of as nocturnal are technically crepuscular: at their most active in the twilight hours around sunset and sunrise. An otter (*Lutra lutra*) is regularly seen at Kirkwall harbour in the Orkney islands in the middle of the night. This individual

Daubenton's bat (*Myotis daubentonii*)

of a usually crepuscular species has found a niche. Kirkwall harbour is brightly lit through the night, providing enough light to make hunting easier for the otter, but at a time when there is little human activity to disturb it.

European beavers (*Castor fiber*) are being reintroduced in Britain. In the daytime, traces of their activity are clearly visible – dam building, gnawed tree trunks and branches, even small saplings felled and branches removed – but it is at dawn and dusk that they work. Beavers are habitat engineers, expanding and maintaining wetlands. Often you will hear rather than see them when they are active, with rustling branch falls and watery splashes.

Daubenton's bats (*Myotis daubentonii*) use waterways as paths to follow and also as hunting grounds, since they feed on insects found just above the water's surface. Bats are equipped with sonar, allowing them to navigate with sound rather than sight. Around the world, nearly all species of bat are night-flying. If they were active in the daytime, they would be on the menu for many day-flying species of birds of prey. For bats, being active after dark reduces their exposure to predators.

Common newts (*Lissotriton vulgaris*), palmate newts (*Lissotriton helveticus*) and great crested newts (*Triturus cristatus*) are nocturnal on land, and hide and rest during the day. They hunt at night, and this is also the time when they travel to reach water for spawning. On land, newts are more active around the new moon, when nights are at their darkest. In ponds during the daytime, newts rise up for a gulp of air and glide back down to hide amongst plants.

At night, rock pools are busy places. Edible crabs (*Cancer pagurus*) use a range of hunting approaches. Mobile prey is stalked and pounced on. Sometimes edible crabs will lurk under rocks and leap out to ambush their prey. They dig up otter clams (*Lutraria lutraria*) from where they are buried at the bottom of sandy rock pools. Despite being voracious hunters and clad in a hard shell, edible crabs' nocturnal activity is thought to be a means of reducing their exposure to larger predators, who can crack their carapace and reach their flesh.

Hermit crabs (*Pagurus bernhardus*) are the reason why shells collected on a beach in daytime can turn into stinking shells that will land in the bin. By day, the shells hermit crabs live in can seem unoccupied from the bird's-eye view of a collector. When turned over, hermit crabs hide in their portable shelter by retreating into their shell and sealing its opening with their large right pincer. Having soft bodies, hermit crabs use hard seashells for protection. As they grow, they move into larger shells. Night-time is when they are most active, with the usual animal pursuit of scavenging for food, and the specifically hermit-crab issue of fighting over shells. Hermit crabs are flexible in their use of periwinkle (*Littorina littorea*) and dog whelk (*Nucella lapillus*) or other shells. However, some shells are more desirable than others. For example, lightweight shells offer an attractive combination of protection with an easier-to-carry burden. Hermit crabs live in aggregations in which they mostly ignore each other apart from mating and fighting over resources of food and shells.

Dog whelks have whorled, pointed shells that are more recognisable for the texture of the ridges you can feel than for the colour – they can be white, grey, brown, yellow, creamy, orangey and combinations of these colours. At night they prowl patches of barnacles (*Balanus perforatus*) and mussels (*Mytilus edulis*). These shells are no defence against dog whelks that can squeeze their proboscis through defensive hard plates or hinging valves. Dog whelks can also drill through shells and create their own entrance for reaching the flesh of the molluscs that they prey on. When

accustomed to light at night, dog whelks spend less time hiding and are more likely to seek prey, even if olfactory cues indicate the proximity of their predators.

Lurking at a rock pool with a torch, you may see opossum shrimp (*Neomysis integer*) fluttering towards the light. They are small, at most about 1.7 centimetres long, making them prey small enough for little cuttlefish (*Sepiola atlantica*) that only reach six centimetres. Because of their small size and their habit of burying themselves in sand so that only their eyes peek out, little cuttlefish are usually hard to spot. Night-time rock-pooling is a chance to see these and other creatures during their active phase.

Fluorescence

A UV torch brings a touch of the tropical to our shores, as here, just like on coral reefs, night-time views brim with fluorescence. In the daytime, snakelocks anemones (*Anemonia viridis*) include some muted grey–beige individuals, and others that have tentacles that are green and then pink-tipped at the ends. It is green snakelocks anemones that stand out bright green under UV light. Brown and white gem anemones (*Aulactinia verrucosa*) blend into rock-pool surroundings in daylight, but under UV light they shine marked in bright green. Anemones, jellyfish and other cnidarians contain green fluorescent protein (GFP). It isn't just a cool optical trick for night-time beach walks; this molecule is used in cell biology and biomedicine. Scientists study function and organisation in living cells and tissues by attaching GFP to objects in order to track them. For example, protein with GFP attached can be watched moving inside cells through a microscope under UV light.

Seaweeds, especially those which are breaking down and decomposing, glow red–orange in UV light as the chlorophyll they contain fluoresces. Seashells and the carapaces of crabs and lobsters light up as their pigments fluoresce. Night-time walks along the strandline of a beach or looking in rock pools with a UV torch are a way of seeing an entire animal or seaweed standing out against its background – in particular because so many seashore animals use pigmentation to camouflage the shape of their body, but UV light reveals the underlying chemical composition that makes living life forms stand out against inert backdrops.

Bioluminescence

In summer's heat, sparkles in the sea are a compelling reason to get immersed – if not for a full swim then at least a toe dipped in sparkling water. Movement in the sea makes sparks flash in eddies and swirls. Perhaps not as epic in the amount of light as tropical bioluminescence, but thrilling nonetheless.

In daylight, the creatures providing this sparkly entertainment are invisible to the naked eye; it takes a microscope to see dinoflagellates (Dinoflagellata). By day, they convert energy from the sun into luciferin. At night, they release light from luciferin in response to movement in the water. This is usually a subtle display of sparks. Occasionally, bioluminescence in water can be visible as a blue glow at the edge of the shore and around swimmers. If you are on a boat, before flushing the toilet, turn off the light; as seawater flushes the toilet bowl, you might see sparkles – a pretty although sorry reminder that away from land, boats can and do empty untreated sewage into the sea.

Bioluminescence occurs in salty or brackish water, not freshwater. Dinoflagellates' flashes of light are thought to disrupt predators grazing on them, and to attract a bigger predator to eat what might be consuming the dinoflagellates. Bearing in mind that the biggest dino-

flagellates are only one millimetre long, swimmers need not worry about what is consuming dinoflagellates. Seeing the sea light up with bioluminescence might take your breath away with excitement, but every other breath you take is thanks to phytoplankton like dinoflagellates. Phytoplankton, or microscopic algae, release about half of the atmospheric oxygen we need to breathe.

Over 1,500 marine species are bioluminescent; on land, only a baker's dozen of insects, a few mushrooms and a scattering of soil-dwelling bacteria illuminate themselves. Bioluminescent displays reach giddy heights in the deep ocean's perpetual night. Predators sport glowing red lures that entice prey to wander in front of their jaws. Sex is on when the lights are on, as fish let potential mates know they are there in the dark via flashing their appendages. The deep ocean is an alien world for us, since the crushing pressure of water at such depths can only be visited by creatures adapted to withstand it. But dinoflagellates aren't the only glowing life forms we can easily encounter on our seashores.

The Romans must have been pleased to discover that their party shellfish, piddock (*Pholas dactylus*), lived in Britain. Pliny the Elder tells us in his *Natural History* how the Romans enjoyed piddocks:

< Fluorescence of animals and seaweeds in rock pools

'It is the nature of these fish to shine in darkness with a bright light when other light is removed, and in proportion to their amount of moisture to glitter both in the mouth of persons masticating them and in their hands, and even on the floor and on their clothes when drops fall from them, making it clear beyond all doubt that their juice possesses a property that we should marvel at even in a solid object.'

Indeed, the Romans enjoyed eating raw piddocks not specifically for their flavour but because they sparkled in the dark and it was entertaining to squash them and be covered in glowing goo. We know the Romans found piddocks in Britain because at a Roman villa in Somerset there are piddock shells in the rubbish heap they left. Piddock-eating carried on into the nineteenth century, but there is no need to eat them to enjoy their glow. They burrow into soft rocks and waterlogged wood and extend a siphon out to suck in seawater. If you see a blue–green light in shallow water or rock pools, it might be from a piddock. Apart from being an example of Roman excess – smashing shellfish on your face because then your face glows – piddocks went on to be a creature from which we learnt chemistry. In the nineteenth century, Raphael Dubois used piddock extract to study bioluminescence. He found out that luciferin and luciferase, an enzyme, were part of the phenomenon. Luciferase has become a useful tool in

science. It can literally illuminate when genes are active, which allows experiments to track cancer cells or detect viruses.

Hawaiian bobtail squid (*Euprymna scolopes*) have led our understanding of the microbiome and human health, as scientists learnt about the process of gut colonisation from their glowing light organs. Bobtail squid have organs that host bioluminescent bacteria. While the bacteria get a home, bobtail squid get a softly glowing underside that makes their silhouette less visible to predators in the sea below them. Our local bobtail squid are the little cuttlefish (*Sepiola atlantica*), which are colonised by *Aliivibrio logei* and *A. fischeri*. In a reflection of our cooler waters, little cuttlefish have more *A. logei* that is more luminescent at temperatures of 20 °C and lower. *A. fischeri*, which has optimal luminescence above 24 °C, is better at colonising Hawaiian bobtail squid.

One of the jellyfish that visits our seas is named for its nocturnal light: *Pelagia noctiluca*, based on the Latin *pelagia*, meaning 'marine', *noctis*, meaning 'night', and *lucis*, 'light'. More pragmatically for daytime, its vernacular name, mauve stinger, is equally descriptive: it is purple and you will feel a sting if its tentacles touch you. Its appearance here is erratic: some years there are no reported sightings, and some years there are so many they

Bioluminescence making the sea sparkle at night >

appear in the newspapers because a swarm has killed farmed fish. Mauve stingers are small enough to fit through the nets that contain fish farmed in the sea and the numbers contained in their swarm are enough to decimate a pen of fish.

Of all our waterside wildlife, bristly marine worms are probably one of the least exciting-sounding. Yet a species that lives in sand, *Caulleriella bioculata*, creates one of the most magical spectacles we can see. A low-tide walk across a beach they are living on at night is like walking on a carpet where you wake up stars, since they glow as your footsteps pass over them.

Perhaps the freedom of long summer days on holiday from school sets us up to associate exploring outdoors with that season. Everyone stampedes to beaches, or riverside parks and meadows, on those sunny days. Yet intertwining patterns of movement of the Earth around the sun create phases of day and season. These induce different behaviours and stages of growth in animals and plants, meaning that autumn, winter and spring offer particular highlights, and night-time hours provide a different view in which some wildlife feels less inhibited. We can appreciate summer and daytime abundance, while consciously making the effort to enter wild places at other times and watching out for the wildlife that marks those times too.

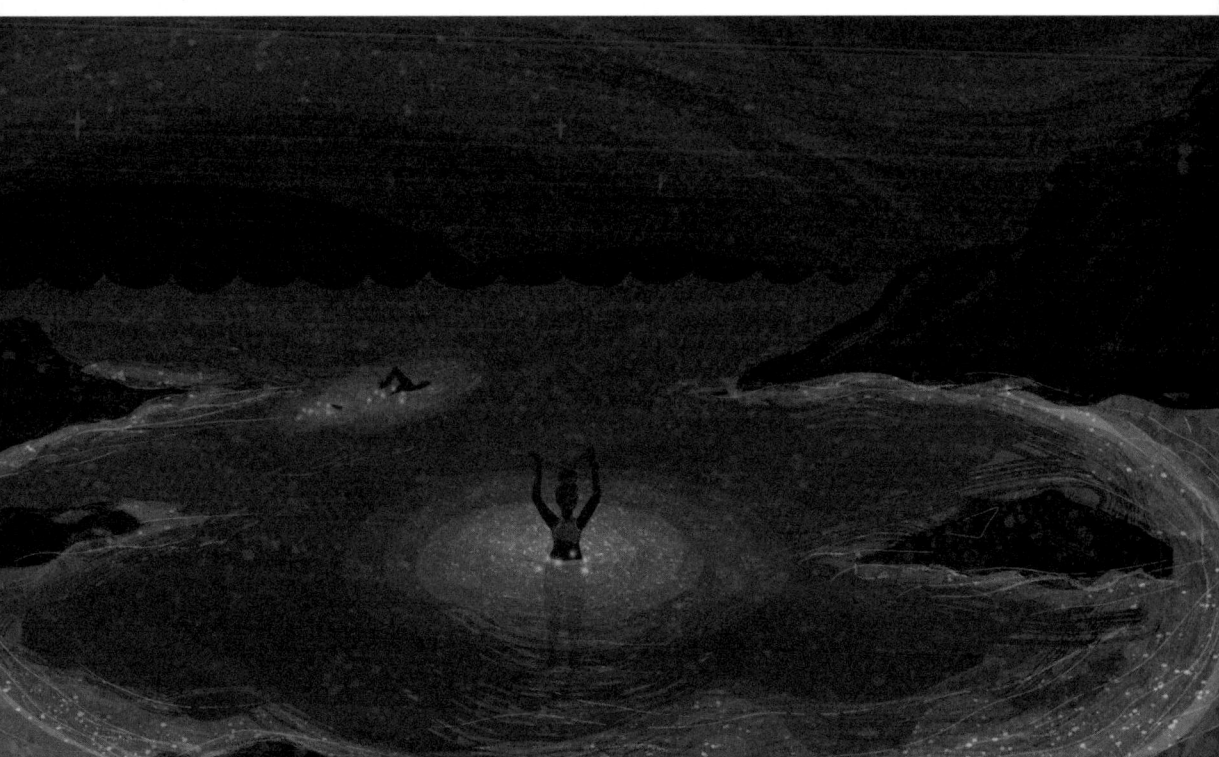

4

Foraging

Pick a pocketful of mint on a walk, or swim for your supper and bring home a blade of seaweed. Foraging isn't like going to a supermarket with produce reliably on shelves for fixed prices. As often as you might choose to collect something growing wild to eat, you might equally choose not to. Perhaps you don't find what you had hoped to, or there isn't as much of it as you thought there would be, or heavy rain earlier makes you think that the run-off from fields containing cowpats into rivers might have temporarily increased *E. coli* levels in the water. When you do collect wild plants and animals, you shouldn't collect everything in sight. Take a little from different plants in

< Harvesting sugar kelp in a sustainable way

different places. A useful guide is that after you have taken some, to a passing person it would not look as if anyone had taken anything. Foraging should include awareness of the environment around you and how other species are engaging with what you are interested in; consciously recognise the ecological web in which you are embedded.

Dr Robin Wall Kimmerer sums up a sustainable reciprocal approach to foraging with Honorable Harvest guidelines. While foraging, we should be accountable for what we take. Not only guided by desire for something, we should also think about who else relies on that resource and think how to harvest it while sustaining it. Always use what you take; if you might not use it, leave it where it is. Don't take the first or last of natural bounty. Not only can early starters or late bloomers be potentially valuable outliers that can adapt to shifting climate, they extend the season of available resources for other species. Share what is yielded to you and look after the species and habitats your harvest comes from.

Foraging and the law

Foraging is covered by aspects of our laws. Under the Wildlife and Countryside Act 1981 you can pick above-ground parts of a plant; however, uprooting a plant without the landowner's permission is illegal. Some plants are considered to be particularly rare and have specific legal protection under Schedule 8 that prohibits picking any parts of them. Under byelaws, some areas that are publicly accessible, such as commons and parks, ban or restrict foraging. You can write to Natural England and request permission to forage on a Site of Special Scientific Interest (SSSI), but we have plenty of publicly accessible land that edible plants and animals live on. There is no need to collect them in SSSIs.

Water quality

Before collecting wild plants and animals to eat, you should be sure that the water they live in is clean. Harbours are a place to avoid in particular. While boats shouldn't discharge toilet water while at harbour, they can discharge waste water from sinks, showers and washing machines. Seaweeds should not be collected for eating from rock pools that are only infrequently refreshed. Likewise, standing freshwater that looks stagnant is not a location to pick aquatic plants. Also avoid places that might infrequently be contaminated, such as downstream from sewage outlets. In theory, sewage is treated before being released, but in practice it can be released without treatment when rainwater exceeds the capacity of sewage systems. Notification of sewage release – if available – can be after the event. Industry is regulated and contaminated water should not be released in our rivers and seas, but it is. While awaiting enforcement of legislation and accountability of polluting industries, it is worth erring on the side of distrust and collecting aquatic foods from locations where you have knowledge of what is upstream. Some contaminants can be dealt with by heat – for example, thorough cooking kills *E. coli*. But chemicals that accumulate in living organisms, such as heavy metals, can't be removed or neutralised prior to consumption. Fish farms use chemical treatments on their stock, and the sea flows unhindered through fish pens. While this has been raised as a health concern for people in the sea near fish pens or using the sea's resources, it is under-researched and not well regulated. Local harvesting has the advantage of being familiar with how habitats should look. You are better placed to notice algal bloom on a lakeshore when you can remember how the water appears most of the year. Being local is also conducive to knowing what is upstream and may contaminate downstream.

In Britain and Ireland, there are more plants and animals that are edible than poisonous ones. Not everything that is technically edible is desirable. Some species are rare or struggling and should be left untasted. Other wild foods are a poor return on investment: they take far more effort finding and preparing than their flavour merits. Fishing is a gateway to sometimes weird but wonderful food. Garfish (*Belone belone*) are a gift to people worried about accidentally swallowing fishbones, since their bones are magnificently obvious, being bright verdigris in colour. However, fishing requires rods, and in most cases licences, and is far more of a planned expedition. Here is a selection of our edible aquatic wildlife, chosen because of the ease with which it can be collected and as an illustration of the range of flavours we can find.

Seaweeds

Occasionally, after a storm has snapped off and washed up fresh seaweed, the strandline can be a place to forage, but in general seaweed washed up on shore is decomposing and not food-grade, although it does make excellent fertiliser for garden plants.

Essential equipment for harvesting seaweed is a knife or pair of scissors. Ripping seaweed fronds off rocks is easily done, but prevents them from re-growing. If you cut off fronds from above their holdfast – which attaches them to rocks – they continue to grow after you have taken some but not all. Seaweed-covered rocks offer bountiful harvest, and also the peril of slipping on seaweed and being injured by landing on rocks. So make sure that your knife and scissors are sheathed while not in use and that you are concentrating on where and how you are walking. Make use of natural grip from rough patches of barnacles. And take your time – treat seaweed-covered rocks like a patch of ice and make sure you have firm footing before taking the next step. Check tide times and aim to collect on an ebbing tide rather than a rising tide, and so avoid being caught out by high tide. Also take weather forecasts into account and stay away from the shoreline when there are storms. Thanks to the popularity of surfing, you can easily check wave conditions online and avoid seaweed foraging when waves are high. Increasingly, no-take zones are being established as marine nature reserves. We have extensive shoreline and much of it is unprotected, so it is easy to enjoy the wildlife we see in no-take zones, and enjoy tasting the fruits of that protection from outside the no-take zones. In marine conservation, a great benefit of the free-flowing nature of the sea is that protected areas can seed other areas with wildlife, including seaweeds which cast their spores out to colonise new areas.

Seaweeds are a forgiving group to begin your foraging journey with, as there are only a few that are poisonous, and these share a distinctive characteristic of very sour taste. Acid weeds (most *Desmarestia*) accumulate sulphuric acid, which probably stops fish and other herbivores eating them. *D. ligulata* and *D. viridis* contain sulphuric acid, and their internal pH can reach 0.5 to 0.8. That is more acidic than your stomach pH of 1.5 to 3.5, or lemon juice, which is about pH 2. Exposed to air, *D. ligulata* and *D. viridis* release sulphuric acid from their vacuoles where it is stored, which breaks them and surrounding seaweeds down. Closely related *D. aculeata* does not contain sulphuric acid but doesn't offer culinary interest, so there is no need to learn how to distinguish it from its relatives. Acid weeds grow in low intertidal and sub-tidal zones. You are less likely to encounter them when foraging on foot, but might if you are swimming or on the sea in a canoe or stand-up paddleboard. Compared

to seaweeds that are edible, acid weeds aren't particularly attractive, looking like a fuzzy tangle of thread. They also taste very sour. If you taste a tiny piece and it is sour, you know it is not a seaweed to eat; while seaweeds span a range of flavours from truffle and salty to sweet, none of the edible ones are sour. As diverse as are the forms that seaweeds take, so are their uses in cooking. Some seaweeds are pleasant to eat fresh collected on the seashore. Others are unwieldy to eat fresh because of their size and texture, and need to be dried, fried or stewed to release their flavour.

Out of the many edible seaweeds, these offer a range of uses, and different textures and colours to enjoy searching for.

Irish moss (*Chondrus crispus*) grows in many colours. In deeper water, it is rich purple–chocolate, and sometimes a blue iridescent gleam flickers on the tips of its fronds. Greenish–yellow forms of Irish moss tend to inhabit shallow water. Identifying this seaweed that wears so many colours is easier using touch. It has flat fronds with blunt tips, and an overall texture that is surprisingly tough and robust. Picking by hand rips the whole plant off the rock; if you use scissors to cut off sections and leave the holdfast that attaches Irish moss to rock, this seaweed can grow back.

Demand for products made using carrageenan extracted from Irish moss increased during the Second World War,

Irish moss (*Chondrus crispus*)

when imports of agar, gum arabica and tragacanth were cut off. Carrageenan was used as a replacement, because it could be harvested along the Atlantic seaboard. On the shore of Prince Edward Island, Nova Scotia, horses were used to pull rakes through the sea's edge to collect Irish moss.

Today the array of consumables available that use carrageenan as an emulsifier, thickener and gelling agent ranges from bread to toothpaste. However, most commercially used carrageenan is now derived from other seaweeds cultivated in Asia. Home-brewing supply shops still sell Irish moss as an effective agent for clarifying beer. A trip to the seashore can easily provide Irish moss for your kitchen cupboard. It is a vegan alternative to gelatine that sets milk or custard puddings to a jellied texture. Irish moss can be used fresh, but most recipes for carrageenan pudding measure it by dry weight. Rinsed and spread out for a few days on a biscuit rack or tea towel, it will dry and can then be stored in a jar. Heating Irish moss in water or milk triggers the release of gelatinous agar, and then the Irish moss is discarded. Five grams of dried Irish moss is enough to gel up to 900 millilitres of fluid. Traditional remedies use Irish moss as a soothing addition in winter teas when coughs and cold air are harsh on throats.

Pepper dulse (*Osmundea pinnatifida*)

Dulse (*Palmaria palmata*)

Sea lettuce (*Ulva*)

Pepper dulse (*Osmundea pinnatifida*) is a peppery taste of the sea. Its colour varies from yellow–green higher up on the shore where it is exposed to more sunlight, to red–brown on the lower shore. However, its frond shape is consistent. Fronds branch alternately and only in one plane – i.e. it is flat like a stencil. Its only depth is the almost succulent thickness of its fronds. A small seaweed, its length from the holdfast at its base to its frond tips is about eight centimetres. Growing attached to rocks in the intertidal zone, pepper dulse is a common and easily found seaweed. Nibbled on the seashore, its seaweed saltiness combined with pepper flavour is a condiment to accompany a walk on a rocky seashore. Other than sampling a taste while out, as a spice in the kitchen it pairs well with fish and cream cheese.

Dulse (*Palmaria palmata*) has fronds that branch off in pairs or hand shapes, and is consistently dark red. These fronds are leathery in texture, and a bit chewy when eaten fresh off the rocks or kelp stipes that it grows on. In cooking, dulse works

Sugar kelp (*Saccharina latissima*)

sautéed with other vegetables, but where it shines is as a dried ingredient which adds concentrated flavour to stews, stocks and soups. Small seaweeds like dulse are easily dried in a warm room indoors, or outdoors on sunny days. When you collect them, rinse them in the sea to wash off sand and grit. Spread them out to dry on a biscuit rack or clean tea towel. This will take several days, during which time you should turn them to make sure they don't settle into clumps. Once fully dried, they can be stored in jars. When powdered, they are easy to add as seasoning, although they retain their flavour longer when left in fronds and crushed or ground as needed.

Sea lettuce (*Ulva*) is the brightest green you will find on the seashore. There are a few species, and the main ones that look like scraps of lettuce and are referred to as sea lettuce are *Ulva lactuca* and *Ulva rigida*. Usually they grow attached to rocks on the upper shore, but if detached from their holdfast in sheltered bays or rock pools, they will float on the surface and continue to grow. Their fronds crumple and are translucent when held up to the light – they are only two cells thick. Their thin texture makes them easy and quick to dry, which also intensifies their flavour. Sea lettuce also works well in a quick pickle of vinegar and sugar, in which it retains a pleasing deep green colour.

Sugar kelp (*Saccharina latissima*) grows as one long frond extending up to four metres long from a short stipe. It is khaki-coloured and has frilly edges. While it might rarely appear in rock pools, sugar kelp tends to be found below the low-tide mark, growing attached to rocks. Collecting sugar kelp is very sustainable if you cut off just the top third section, leaving most of the frond still attached to its stipe. Like kale, if spritzed with oil and toasted in an oven, this seaweed makes great savoury crisps. Long lengths of fresh sugar kelp are flexible enough to wrap around parcels of whole fish or meat for oven baking or cooking over a slow barbecue. It adds flavour as well as retaining moisture in the fish or meat through the cooking process. These big lengths of seaweed are easily dried hung over a clothes rack or plate drainer. As they dry, white powder appears on the surface – not mould but mannitol. Taste it and you will understand why this kelp is named for sugar. Dried pieces of sugar kelp keep well in a jar and are perfect for dropping into a stockpot or stew for adding flavour.

Leafy greens

Making a foray into leafy greens, one plant you must learn to identify and must not eat is water hemlock (*Oenanthe crocata*), our most poisonous plant. It grows in abundant lushness along freshwater. Deceitfully, its leaves look like benign and flavourful flat parsley. But water hemlock isn't just a plant that can make people sick; it is lethal. Oenanthotoxin – a convulsant – is in its leaves, stems and tubers. Symptoms of poisoning by water hemlock range from vomiting to hallucinations, tachycardia and seizures. Get to know this plant so that you can be sure to avoid consuming it. Water hemlock's guise of familiar leafy herb and appealing-looking tubers is a reminder to operate within the bounds of your knowledge and to resist being tempted by plants that look similar to known edibles if you don't know the identifying features that you are looking for. Cases of water-hemlock poisoning in Britain and Ireland have involved failing to identify the plant as water hemlock, and consuming it thinking it looks like a different herb or vegetable, such as parsley or water parsnip.

Foragers must also beware of cowbane or northern water hemlock (*Cicuta virosa*), a scarce but poisonous plant that contains fast-acting cicutoxin. Both cowbane and water hemlock are members of the carrot family – Apiaceae. This group of plants includes both eminently edible members and highly poisonous members. For this reason, people beginning to forage wild plants might usefully spend at least a year watching carrot family members and getting to know them in their different stages of growth. Their flowers are distinctively arranged in umbels – like an umbrella from which the fabric has been removed and each rib of the umbrella topped with a flower. Carrot family seed heads also hold the umbel structure and often remain upstanding through autumn and winter.

In spring, their pinnate leaves – leaflets in pairs along the leafstalk – emerge. Leaves are followed by flower stems, which bear useful identification characteristics of hairs and blotches.

Sea beet (*Beta vulgaris*) has shiny, succulent leaves in late winter and spring. Jostling against each other, they squeak. As summer progresses, the leaves at its base become larger and worn, and the leaves on its flowering stem are progressively smaller further up the stem and become too fiddly to pick. In any case, despite being in leaf nearly all year round, sea beet is best for eating when the growth is fresh in winter and spring. Treat these leaves like spinach, but as they are sturdier they don't have spinach's disappointing ability to shrivel to a tenth of its size on being cooked.

Watercress (*Nasturtium officinale*) grows in freshwater on calcareous substrates; you will see it in streams, rivers, canals, ditches, ponds and marshes. Sometimes it even grows above the strandline on beaches where freshwater forms pools and streams. Bought in a shop, you can enjoy watercress raw. But wild watercress can only be eaten raw if you are 100 per cent certain that the leaves have not been submersed in water. This means that leaves must be picked well above the height of the water, and that the height at which they are picked must be above any fluctuations that might have occurred in water levels. This is because watercress is the host plant for sheep liver fluke (*Fasciola hepatica*). Usually this worm, which has an intermediate stage as cysts on watercress, infects sheep. But if people consume raw leaves with cysts, sheep liver flukes will inhabit human livers instead of their usual sheep host. We can enjoy watercress when it is cooked, as the heat of cooking kills the immature cysts if they are present on the leaves.

Marsh samphire (*Salicornia*) is a succulent salt-marsh plant composed of branching tubes. There are seven species of marsh samphire known to grow around England and Ireland, but they are tricky to identify, to the extent that official records of their distribution are regarded as provisional. They are all edible, but some species are more common than others. So, as always, forage with consideration for individual plants being able to continue growing after you have taken some. And remember that the patch of marsh samphire should look as abundant after you have made your harvest as it did when you arrived. Although marsh samphire looks like it will easily snap, its core is tough and fibrous. It must be collected using a knife or scissors to take sections from the juicy tips. Trying to pull up marsh samphire either results in uprooting the whole plant or uprooting you as the plant remains standing and you crash-land on the mud. Sautéed in butter, marsh samphire is delicious; less indulgently, it is also good briefly boiled and drained. Treat marsh samphire stalks like tiny, pre-salted asparagus tips and you will enjoy them.

Sea sandwort (*Honckenya peploides*) forms bright green mounds at the top of sandy beaches. It doesn't tolerate being immersed in seawater, but it does withstand shifting sands. If you lie down for closer appreciation, you can brush your hand over it and take a moment to enjoy a gentle scent of cucumber being released. Best eaten before it has flowered, you can eat sea sandwort's young shoots raw or cooked. Take time to brine and lacto-ferment them and you will be well rewarded with a deliciously moreish pickle.

Sea purslane (*Halimione portulacoides*) has to be one of the easiest plants to forage. It is a grey-leaved shrub that is in leaf all year and grows on salt marshes and rocky shores. Nothing is easier than picking a small branch as you walk past on your way home. Strip the leaves off by running stalks through your fingers against the upwards direction in which the leaves point. Scatter these leaves over anything you want to roast, add a splash of oil and whatever you have roasted will be flecked with crispy, briny leaves. This might be the quickest way to bring a taste of the sea home.

Aromatics

Offering concentrated aroma and flavour, some plants are only needed in small quantities. The master of distilling plant flavours is our decimated and now reintroduced European beaver (*Castor fiber*). We didn't hunt them to local extinction just for their pelts; beavers were also the source of castoreum, which they secrete to mark their territory, the aromatic quality being a barometer of how good they are at finding food. People used castoreum in perfume and to flavour ice cream, in particular in strawberry-flavoured ice cream. Instead of using the sweat equity of beavers' hard work, we can find intense aromas in some of our waterside plants.

Rock samphire (*Crithmum maritimum*) is like a savoury sweetie with no shopping or unwrapping required – just snap off a leaf and chew it for a burst of flavour as you walk on rocky seashores. Its habit of growing from between cracks in rocks means that is often above dog-weeing height, so even on popular seashore paths you can eat it straight off the plant. Rock samphire has pointed evergreen leaves, which in the classic style of plants adapted to a lack of freshwater are glaucous grey. The colour is due to a waxy skin that reduces water loss. Underneath, the leaves are fleshy as they store water. A member of the carrot family, its flavour is a concentration of carrot, like freshly crushed carrot leaves magnified. You can use it fresh as a seasoning. Being evergreen, it is available year round, and its fresh flavour is better than when dried. Alternatively, bottle rock samphire leaves in vinegar with salt and a clove of garlic. This matures into aromatic pickled leaves, but don't discard the vinegar when you eat the leaves. Use the vinegar in sour glazes and salad dressings.

Bog myrtle (*Myrica gale*) is a stealthy plant that looks forgettable until you brush against it and catch the rich, resinous smell it releases from its leaves and flowers, a bit like rosemary and frankincense combined. Bog myrtle grows in knee- to waist-high thickets on bogs, marshes, fens and wet heathlands. It was the dominant additive in beer until the thirteenth century, when the use of hops as a bittering agent became more common and eventually dominant. Its inconspicuous flowers borne on bare twigs in spring are concentrated with flavour, more fiddly to pick than the leaves that appear after flowering. Both leaves and flowers can be dried or frozen for later use. The easiest way to enjoy its flavour is to drop the leaves in a bottle of vodka or schnapps. When it becomes golden brown, you know that bog-myrtle flavour has infused into the liquid. Discard the leaves or flowers and enjoy it as shots Danish-style – in Denmark this is known as *porse aquavit* – or blended into longer cocktails.

Watermint (*Mentha aquatica*) is a more fragrant relative of spearmint (*Mentha spicata*), a garden plant. While spearmint is just minty, watermint has perfumed notes of eau de cologne. In leaf, watermint might be confused with cornmint (*Mentha arvensis*), which grows in damp soil, not in water. When flowering, you can see that watermint stalks are topped by flowers, whereas cornmint stalks have flowers topped by a tuft of leaves. Both watermint and cornmint are edible. Pennyroyal (*Mentha pulegium*) is a rare plant found in damp grassland, particularly where there is seasonal inundation. Pennyroyal is hepatoxic (damaging to liver cells) as it contains pulegone, a naturally occurring monoterpene. As a rarity, you are unlikely to encounter it, and if you stick to collecting watermint by ponds and riverbanks you should avoid accidentally picking pennyroyal. Watermint reaches into water and emerges growing out of water, so pick it high on riverbanks where you can take leaves that have not been submersed. There is some evidence from Central Asia that sheep liver fluke cysts, although predominantly hosted by watercress, may use some mint species as hosts. But don't be deterred from enjoying watermint – the risk of sheep liver flukes on it is low. Since watermint grows on riverbanks and at lakesides as well as in streams of flowing water, it can easily be picked from sites where leaves have not been immersed. Watermint is best used fresh, as drying loses some of its aroma. It makes good minty tea, and is excellent in cocktails from mint julep to mojito where its eau-de-cologne element differentiates it from cocktails made with standard spearmint.

Meadowsweet (*Filipendula ulmaria*) is found on the margins of waterways, lakeshores and damp ditches. Its creamy white flowers make it stand out in summer, but get to know how to recognise its leaves, which are aromatic in spring, and you can collect the sweet flavour of this plant for a longer season. Both leaves and flowers contain coumarin, a hay-meadow and vanilla scent that enhances many drinks and puddings. Meadowsweet also contains other chemicals that add almond and wintergreen flavours. Leaves and flowers can both be used fresh or dried. When drying meadowsweet, it is important to put it on a rack where air can circulate under and over it as soon as possible after collection. Make sure it is well spread out in a single layer. Delicious coumarin can be converted to dicoumarol, which is toxic, by some moulds. So meadowsweet needs to be dried efficiently, and any batches that develop mould should be discarded. Fresh or dried meadowsweet makes beautifully fragrant tea. It also works well infused into sorbets and cocktails. In hot water or

lightly heated grape juice, meadowsweet's delicate flowers are quick to yield their flavour, but tougher leaves are worth leaving to stand for twenty minutes or a few hours to get the flavour out.

Crustaceans

Lobsters, crabs, prawns and dainty shrimps share characteristics and ancestors. As a group, crustaceans are invertebrates that have segmented bodies, and legs in pairs. Predominantly, crustaceans are aquatic, although there are some exceptions like woodlice (Armadillidiidae). Crustaceans, as their name suggests, have a rigid external crust or body shell that is moulted and shed as the animal grows.

Animal welfare laws in the UK don't currently apply to crustaceans, although pressure is growing for these edible animals to be recognised by welfare laws. This means that despite proof that lobsters, crabs and prawns are sentient and do feel pain, the way they are brought to restaurants and shops does not use methods of storage, handling and slaughter that are humane. Between being caught and being killed, they are not insensible to distress and pain. You might argue that catching these creatures yourself, killing them soon after catching them, and eating them fresh at home is a more humane approach.

European lobster (*Homarus gammarus*)

European lobsters (*Homarus gammarus*) are one of the more expensive items you'll see on a restaurant menu, but if you dawdle on a swim or dive you can see them in the holes and crevices they lurk in underwater. Compare this habit to how you see them kept alive in supermarkets under bright lights. It is a long torment for lobsters in shops between being caught and being cooked. If collecting lobsters, you are a hobby fisherman. As such, you may not sell what you collect, as this would be commercial fishing that requires a permit. There are regional variations in the maximum catch allowed: in England and Wales, two per day; in Scotland, one. Additionally, lobsters have a minimum catch size: if their carapace length is less than ninety millimetres, they are considered too small to have bred and need to be left in the wild breeding pool, not taken for a cooking pot. If you find a lobster with lots of small jelly-like spheres attached, this is a female carrying eggs. Along with her clutch of eggs, she should be left in the sea. Likewise, lobsters with a V-shaped notch on their tails should be left in the sea, as they have been marked as known breeding females. These regulations by inshore fisheries stakeholders help to maintain wild lobster populations as a sustainable fishery. We can eat lobsters, but to eat lobsters in the future as well means protecting a viable lobster population now. If you pick up a lobster it will try to defend itself, so watch out for its claws, especially the one adapted for crushing shells we would need a hammer to get into. That claw can easily break a finger.

Boiling lobsters alive is not a humane way to treat them. Because they have a decentralised neural system, they cannot be quickly dispatched by stabbing them at one point. Currently the way to kill lobsters that is considered to be the quickest and least painful is to first stun them. At a domestic level, this can be done by chilling them in a freezer for at least twenty minutes. You will know they are insensible when they have no resistance to handling, no eye reactions when their shell is tapped, and no reaction when touched around their mouthparts. Once stunned, they can be killed by splitting them along the midline of their underside, which also destroys their nervous system. Treating lobsters

humanely before you eat takes more effort, but if the taste of them is not worth the effort, just leave them in the sea.

When you see a wild lobster, you might wonder why they are blue when all cooked lobsters are red. Red pigment absorbed from their diet is bonded to protein in their shell and appears blue. Heat damages protein, so in the course of cooking, protein's influence on colour is negated, leaving the stable red pigment.

Crabs (Brachyura) are a group of crustaceans around our shores that includes several edible species.

Spider crabs (*Maja brachydactyla*) are usually caught in crab pots in deep water. But you can see them on beaches too. They have spines on their shell and long, lean legs. With bodies up to twenty centimetres long, and leg spans of fifty centimetres and wider, they are our largest crab. The minimum size at which they should be kept if caught is when their body is thirteen centimetres long. They are easy to catch if you pick them up from behind, where their pincers cannot grab your hand.

Edible crabs (*Cancer pagurus*) are another large crab, with a body that can be twenty-five centimetres wide and a minimum catch size of thirteen centimetres. Their pincers have black tips, in contrast to their red–brown bodies and legs. They can be found in rock pools and

Clockwise from top left: velvet crab (*Necora puber*), edible crab (*Cancer pagurus*), shore crab (*Carcinus maenas*) and spider crab (*Maja brachydactyla*)

shallow waters, and can be caught by picking them up if you are deft. You can also set up a crab pot with bait and check it six to eight hours later to see what you have caught.

Velvet crabs (*Necora puber*) are smaller at eight centimetres wide, and are quick swimmers that you would struggle to catch by hand. Their minimum catch size is 6.5 centimetres. Shore crabs (*Carcinus maenas*) are the most common crabs along our coasts and are similarly sized. These smaller crabs can be caught by dropping a net with bait in it into

the sea on a line. Looking down into the water, you'll see when a crab is eating the bait through the net. If you don't have a clear bird's-eye view, you will need to rely on feeling the line vibrate as a crab plucks at the net.

Velvet crabs and shore crabs are fiddly to get crabmeat from, and are best used in recipes like bisque where their flavour is extracted without much effort. Spider crabs and edible crabs are big enough to be worth the work of separating crabmeat from shell. In all cases, you should check that they are at least minimum-catch-sized and are not berried (egg-carrying) females before taking them away from where you caught them. Like lobsters, they should be chilled to insensibility in a freezer before you kill them. Crabs have two nerve centres, both in the middle of their body. One is towards the front, below a shallow depression; the other is towards the rear, just above the end of where their tail folds under. They can quickly be killed by using a sharp knife in their underside to impale the two nerve centres.

Shellfish

Around the world, the durability of shells has been mined for information on how our ancestors lived. Shell middens, rubbish heaps of discarded shells, clearly indicate the presence of people. Shells can be dated to find out when people left them, the kinds of shellfish they ate can be identified, and some trace analyses can suggest how they might have been prepared for eating. Each shell midden tells a story of the food that people ate, but the details vary from place to place. What is clear is that shellfish, being easily and reliably collected, have been an important food for human cultures.

Bivalves (Bivalvia) are invertebrates that compress their bodies inside a hinged shell of two parts. While inside the shell, its durability protects them from many predators. However, in order to feed, move and breed, they must expose themselves by opening their shells. Predominantly filter-feeders – collecting their food by sieving water passing by – they can concentrate pollutants in the water. When collecting bivalves for food, the first requirement is that the sea must be clean. Also, any bivalves that smell bad at any stage of collecting and cooking should not be eaten. They are available and can be eaten year round, but as seawater cools, microbial growth slows, and the saying to only eat shellfish when there is an 'r' in the month is a reflection of how harmful algal blooms are more likely at warmer temperatures. During algal blooms, toxins released by algae can accumulate in filter-feeders and cause shellfish poisoning in people who eat them. Purging bivalves by leaving them to sit in a bucket of tap water in which thirty-five grams of salt has been added per litre of water can clean them by removing or significantly reducing bacteria and algal toxins in them. This also reduces the amount of grit and sand they will contain when you eat them. Purging isn't considered effective for removing any live viruses they may be hosting. Quick steaming does not remove the risk of viral poisoning, as the temperatures reached inside the shell might not be hot

Mussels (*Mytilus edulis*) growing on a buoy rope >

enough to kill viruses. However, thorough cooking above 60 °C does. Bivalves' ability to open and close their shells provides useful cues to foragers on their healthiness and suitability for consumption. After being collected, bivalves should be tightly shut – the living animal having closed the two halves. After cooking, when the animal inside is killed, the shell should be open. Any shells that don't follow these behaviours should be discarded as unsuitable for consumption.

Mussels (*Mytilus edulis*) spend their adult life fixed to a substrate, which can be seabed, rocks, pier legs or buoy ropes. This sedentary life makes them easy to find and return to. At low tides, some mussels are exposed above sea level and are easy to reach, but mussels that stay below the sea at all tides tend to be bigger. Don't waste time collecting small mussels; stick to those which are five centimetres or longer. You can grasp and twist mussels to pull them off their point of attachment, but you might want a knife to chip off barnacles and epiphytes attached to their shells.

Razor clams (*Ensis*)

Take them home to purge their insides of grit and sand, and reduce or remove bacteria they may contain, in a bucket of salted water. An easy way to cook them that ensures there has been enough heat to kill any viruses is to turn them into *moules marinière*.

Razor clams (*Ensis*) are shellfish that involve a tug of war to catch each one. There are several species of razor clam around Britain and Ireland, including *E. arcuatus*, *E. ensis*, *E. magnus* and *E. siliqua*. All of them are delicious and succulent inside their long rectangular shells, and live buried in sand and silt. There isn't a selective way for commercial fisheries to collect them. Dredging and electrofishing can pull them up, along with all the other creatures living in the sand. Perhaps this makes razor clams the best shellfish to seek out, as you can't buy them. Hand-collecting them has low impact, other than for the individuals

you catch. Check tide times and find a low spring tide for heading to a sheltered bay where you have seen their shells washed ashore and the sand exposed at low tide is firm. A tub of salt is all the equipment you need for catching razor clams. Tasting better than scallops – great scallops (*Pecten maximus*) and queen scallops (*Aequipecten opercularis*) – which need a free diver or scuba diver to collect them from the seabed, razor clams and their low-tech collection punch above their weight. Walk out barefoot or in wellington boots and look for holes in the sand that look like keyholes. When you find one, trickle salt in and wait. If the sand is dry, pour a little water on to wash the salt down the keyhole. Wet sand will be pushed out, followed by a shell rising up from the sand. Let it emerge so that enough length is exposed for you to grasp it firmly with your hand and pull it entirely out. Hold tight, as the razor

Common cockles (*Cerastoderma edule*)

clam will try to burrow back down. You can't tell how large each individual will be until you have caught it. At the end of your razor-clam hunting, return those that seem small to the sand in the area you caught them so they can burrow back down and grow older. After purging the razor clams you keep, lightly steam them so that the shells pop open. Remove the dark-coloured flesh, which is their stomach, and then treat the pale flesh like squid. If you could buy them in shops, they would taste delicious; but as a really wild food that has to be gathered by hand, they carry an extra frisson of delight.

Common cockles (*Cerastoderma edule*) live in the upper layer of sediment in the middle-to-low intertidal zone of beaches, so collecting time is at low tide.

Sometimes their shells are visible on the surface. Even when buried, there is usually a mark on the sand suggesting something is beneath. Common cockles don't flee fast and can be lifted out of the sand by hand, trowel or rake. They don't take long to collect. After being purged, boil them so that the shells open and the meat is exposed to boiling water. The British style is to eat them with a splash of vinegar. Or follow Italian recipes for pasta *vongole*, in which the cockles are heated in a covered pan until they open. Then remove the meat from the shells and sauté with oil, garlic and chilli for a few minutes, followed with a splash of wine that is cooked off. Finish with chopped parsley and serve over pasta.

5

Catching Memories

Pinning down tangible scraps of experience outdoors anchors the intangible. Keeping records is a human impulse we can see as far back as ancient rock art. Record-keeping takes many forms, as drawings, writing, traditional storytelling and collections of objects. Over time, whether daily or sporadically made, records become more than a sum of entries. Themes emerge: some places and times might be stories of seaweed, others counting birds. Comparing present to past, or differences between older and more recent records, gathers predictive power. Seeing or experiencing something again can be planned according to when it is likely to happen

< Seaweed diversity takes many forms and textures (*see page 177*)

in that place. Individual records aggregated together build collective memory of wildlife and the space it inhabits. Feeding into national and international wildlife databases forms data sets that can inform development planning and conservation decisions that subsequently shape the future.

Colours

On a remote voyage that shaped our world – with evidence of evolution and the collection of blight-resistant potato seeds – Charles Darwin used a system to describe colour. On 7 January 1832 he wrote, 'clouds, varying in tint between a hyacinth red and a chestnut brown, were continually passing over the body'. Darwin was describing the colour of a cuttlefish using

Werner's nomenclature of colour adapted by Patrick Syme. Abraham Gottlob Werner was a geologist who set out a language of colour parsed in minerals. In mineralogical terms, hyacinth refers to the brown crystal now known as jacinth, not the bright blue, pink or white sweetly scented flowers you might first think of. Patrick Syme added the living world as a frame of reference to Werner's nomenclature, and Darwin took Syme's book on the voyage of the *Beagle* so that he could record his observations with accuracy. Syme included additional notes and colour blends in the book, so that Prussian blue is the colour of 'Beauty Spot on Wing of Mallard Drake', emerald green is 'Beauty Spot on Wing of Teal Drake', leek green is 'Sea Kale', wax yellow is 'Larva of Large Water Beetle', and so the concept of colour is communicated in animals and plants. In the world of print, we have moved to Pantone Matching System (PMS) and Cyan, Magenta, Yellow and Key (CMYK). On electronic devices, colour is specified in Red Green Blue (RGB) and HEX, a hexadecimal code of numbers and letters.

Building your own colour code using wildlife that you see at a time or place might lead you to a deeper understanding. For example, at Dungeness in Kent, Derek Jarman's Prospect Cottage has yellow window and door frames. Once you've seen clumps of gorse flowering by the cottage, you might think of that shade of yellow as gorse yellow. Under dull winter skies, in a superficial glance across the shingle of Dungeness towards a grey–brown chunder of sea, colours seem limited. Looking closer reaps colourful rewards. On stones and branches, patches of orange stand out – maritime sunburst lichen (*Xanthoria parietina*). An acid-green mound is broom fork-moss (*Dicranum scoparium*). Common greenshield lichen (*Flavoparmelia caperata*) is celadon for now; in summer conditions it looks greyer. Here is a palette of Dungeness in winter, which reflects the site's rare beauty. Dungeness hosts a scarce vegetation type of lichen-rich shingle heath. In summer the colours are different: horned poppy (*Glaucium flavum*) yellow and viper's bugloss (*Echium vulgare*) blue, along with other seashore plants, form a palette of vegetated shingle, another rare habitat.

Blending paints to match a colour you are looking at can be time-consuming. A shortcut is to use the ratios described by Syme and then adjust them to what you see. Or make use of modern technology and take a photo. Digital artists can use an online colour picker to match a selected colour from an image with its RGB or HEX code. Whether you draw and write a colour chart like Syme, or curate a collection of photos, highlighting specific colours catches a place and the feel of it at a certain time.

Charcoal

Most art supply shops sell materials made from a waterside tree. Willow charcoal is still in demand and is made from twigs of willow (*Salix*). Compressed charcoal and charcoal pencils haven't replaced artists' appreciation for willow charcoal as a drawing tool. It is easy to erase because it does not contain binders. This makes willow charcoal a forgiving medium for beginners to work with, and makes creating highlights (pale areas) easier, as marks can be modified or removed.

Wood that is heated but prevented from being consumed by flames turns into carbonised wood. Making artists' charcoal uses willow twigs, an unpainted metal tin with a lid, a fire and a pair of tongs. Young willow twigs of a year's growth work well. First, make a couple of holes in the metal tin so that gases can escape as it is heated. Then cut your twigs into lengths that fit in the tin. After peeling the thin bark off the twigs, leave them to dry overnight. Then fill the tin with twigs, shut the lid and use tongs to put it in a fire. The amount of time the process takes will vary in relation to how hot the fire is, the ambient temperature and how much wood is in the tin. Half an hour to an hour of heating should be enough. Remove the tin from the fire – but before opening, the tin and its contents must be left to cool. If opened while still hot, being exposed to oxygen

in the air will allow the willow to ignite. When ready for use as charcoal, the willow twigs will have shrunk and become black. If the twigs need further heating, close the tin and put it back in the fire a little longer.

Sepia ink

Common cuttlefish (*Sepia officinalis*) are an enticing prospect for ink-making because of their long history of being used as sepia ink, as far back as Ancient Greece. Georges Cuvier, a French scientist of the early nineteenth century, illustrated his *Anatomy of the Mollusca* with drawings created using ink he collected while dissecting cephalopods – squid, octopus and cuttlefish, all of which belong to the mollusc group of animals. In 1833 Elizabeth Philpot wrote to Mary Buckland and enclosed a drawing that brought 200-million-year-old creatures to life. She drew an ichthyosaur head using ink from a fossilised squid of the same age. In 1875 Henry Lee, naturalist at the Brighton Aquarium, wrote about fishermen who dried the ink sacs of cuttlefish and sold them to agents of Mssrs. Newman's, a London-based artist's supplier. Dried ink sacs were boiled in a solution of soda or potash to extract the pigment. The liquid was filtered and then precipitated with acid, filtered, washed and dried. Artists

Common cuttlefish (*Sepia officinalis*)

would then combine the dry pigment with water and gum arabic.

In 2015, while artist in residence at the Grant Museum of Zoology in London, Eleanor Morgan used fresh cuttlefish ink to make prints of fish. She had been told about *gyotaku*, a traditional technique used in Japan to record the size of fish caught by using their bodies as printing plates. In the simplest form, the fish served as a printing plate. Damp rice paper was laid over an inked fish and pressed to ensure full contact. The paper could then be

lifted off with the fish marked on it. More advanced indirect and transfer methods were also used to ink impressions of fish on paper or cloth.

If you don't feel like dissecting a cuttlefish to get ink, you can buy it from cooking supply stores, where its intended use is in risottos and pasta. If you get hold of fresh cuttlefish ink sacs, it might be easier to use the ink fresh, instead of drying the sacs and having to dissolve pigment, remove squid tissue, precipitate pigment and then dilute with carrier fluid for use.

Mapping

Even maps that present as objective contain systemic bias, because of the challenge of representing three-dimensional space in two dimensions. We are most familiar with a way of projecting the world invented by Gerardus Mercator in 1569. While ideal for navigation at sea, the Mercator system relatively increases the size of land at the poles. While less obvious in the southern hemisphere, because there is more ocean and less land, in the northern hemisphere the distortion is clear. Greenland is in fact not the size of the African continent, as it appears in a Mercator projection; rather it is about the size of the Democratic Republic of the Congo. Maps of the London Underground and many other city transport networks are schematic, showing relative position but not geographical locations.

Knowing that map-making is driven by intended function and not bound to accurately represent geographical space is liberating – it affirms that if you make your own map of an area, the guiding criteria are the ones that you set, for example, a schematic map in which you connect places you like to go to by the transport corridors you follow to get there. This might group locations on the same road closer together. Or a map centred on your home might include locations arrayed by the time it takes for you to reach them. Places that are close as the crow flies might be represented as further away because of the distance you have to drive over roads to reach them. You might focus on a single watercourse and over a day, a week, months or years fill in around it what you see when you go there. Perhaps you particularly enjoy seeing otters and want to bring together places where you have connected with them. This might include spots where you have seen otters hunting in the sea, freshwater pools – which seagoing otters need to clean their fur of salt – and sleek patches of mud that are otter slides. Also, note places where you have found otter poo, which distinctively contains fish scales and smells of jasmine.

So first pick your perspective, and list sites you want to mark. You might refer to a conventional map to plot your sites geographically or let your mind shape the space between them. Once your sites are marked, add a small drawing that illustrates each one for you: willow tree, waterfall, ocean wave and other elements can all be represented with little icons you can create. Then add landscape features you notice, such as a shoreline, river's course, cycle path or road. You can add a compass with a north heading, pointing to your home, a mountain or any feature off-map that you want to anchor the map to. Add a key in the corner of the map explaining the symbols you have used. Leave your map sparse and lean, or cram in markings of objects to fill in all the blank spaces. It might turn into a map you keep adding to, or you might trace its bones and repeat with different features picked out at different seasons. Map-making creates a receptacle for your memories and future plans. It is also a process in which you engage with yourself – what grasps your attention – in a place.

Marbling paper

In the seventeenth century, people in England were aware of the art of marbling paper. Sir Thomas Herbert wrote in *Travels in Persia, 1627–1629* that 'paper of a curious gloss and fineness varied into several fancies, effected by taking oiled colours and dropping them severally upon water, whereby the paper becomes sleek and chamleted or veined, in such sort as it resembles agate or porphyry'. Known in the Ottoman Empire and in contemporary Turkey as *ebru* (art of clouds), marbled paper was initially called Turkish paper in England. It was an exotic product as the art of making it was unclear, and even when Europeans understood the process, one of the critical ingredients was not as readily available in Europe. The discovery that the mucilaginous texture derived from Irish moss (*Chondrus crispus*) could be used to treat water to make paint float on it – instead of the gum tragacanth

and orchid tubers that were used in the Ottoman Empire – catalysed the production of marbled paper in Europe. Marbling as a commercial art wasn't just floating paint on the surface but controlling its distribution in order to create and replicate patterns, using tools to push and pull the paint. From the mid-seventeenth century until the nineteenth century, marbled paper was in demand as endpapers in books. In the nineteenth century, books became cheaper and less decorated, and the use of marbled paper declined.

To marble paper, first prepare a bath of thickened water for the paint to float on. Add room-temperature water to gelatinous water drained from Irish moss that has been heated in water, until you get a consistency of wallpaper paste. Carrageenan content is variable in Irish moss, depending on its age and the environmental conditions it has grown in. Taking the guesswork out of preparing the sizing (thickened water) is easily done if you buy powdered 'lambda' carrageenan and dissolve five grams in three litres of water. 'Kappa' and 'iota' carrageenan are extracted from different species of seaweed and will form gel that is too thick. Let the size sit for six to eight hours so that any air bubbles made during mixing dissipate. Treating paper with mordant ensures that paint will stay on it. Alum (aluminium potassium sulphate) is the mordant used in marbling. Dissolve ten grams of alum in 120 millilitres of boiled water and paint it over the paper you want to use about twenty minutes before marbling. In order to keep paint separate, it is mixed with surfactant. Traditionally, ox-gall was used, and this is still for sale in art supply shops. If you prefer an alternative, washing-up liquid works, although not quite as well as ox-gall. Paint that is dropped, using a paintbrush or pipette, on to the size will spread over the surface. The simplest manipulation is to drag a point through different colours of paint, causing them to swirl into each other – just like marbling icing on a cake. When you have an assemblage of colour and pattern you like, lower a piece of paper towards the surface. Its middle should touch first and then allow the sides to descend. Use the short ends to lift it out of the water, the way you would turn a page on a book if you were using two hands to turn the page. Leave the paper paint-side-up to dry. Once it has dried, rinse it to remove traces of size; the mordant will hold the paint to the paper, allowing the carrageenan to be washed off.

Phytotypes

Sir John Herschel explored light-sensitive material at the dawn of photography – not just the silver-based black and white and sepia process, but also different colour contrasts. He invented blue cyanotypes that use iron, and purple chrysotypes that use gold. He also looked outside of light-sensitive metallic salts and invented a way to create photos using plant pigments. Extracts from plants were used to dye paper, and then selective exposure to sunlight bleached areas. He thought phytotypes were not viable because the process of making them was slow and the results were not durable, as fixatives did not prevent fading. Today this ephemeral art can be given permanence by scanning or taking a photo of the end result. Phytotypes appeal if you are concerned about using mordants and metallic salts that are then rinsed into our waterways. In addition to plant material, all that is needed to make a phytotype is water or alcohol as a solvent, a pestle and mortar or blender, and a frameless glass photo frame.

First, select a plant or seaweed to extract pigment from. Beach rose (*Rosa rugosa*) petals are hotly pink and smell beautiful to work with. A bonus is that this introduced species is not appreciated for its enthusiastic colonisation of seashores, which in some areas is considered detrimental to other species. Making use of its abundance is a friendly alternative to trying to eradicate it. Meadowsweet (*Filipendula ulmaria*) leaves and stems are rich in water-soluble tannins. Green seaweeds and red seaweeds are dense in chlorophyll and phycoerythrin respectively. All of these are suitable for human consumption and can be processed using kitchen equipment.

Combine your plant material and grind or blend it with a little water or alcohol. Filter it and then paint paper with the liquid. Leave the paper to dry in the dark. Select plant material, or other objects with interesting outlines, to use as the subject of your image. If you look at watercress (*Nasturtium officinale*) when it is flowering, you'll notice that the leaves at the base of the plant are rounded and the leaves on the flowering stalk are pointed. A leaf from both the base and the crown of watercress makes an intriguing diptych. Similarly, watercress and fool's watercress (*Helosciadium nodiflorum*) look superficially similar when seen growing in streams and rivers, as they both have pinnate leaves – leaflets in pairs on opposite sides of the leaf stem. Fool's watercress leaves are pointed at the tips, unlike the basal leaves on watercress but similar to their upper leaves. A phytotype of fool's watercress alongside both types of watercress leaf creates a triptych that illustrates how fool's watercress is marked by more toothed (jagged) edges to its leaves.

Take a dry sheet of your tinted paper, put it on an open photo frame and press a leaf, piece of seaweed or other object flat on it. Fresh plant leaves can be springy and unobliging in being arranged, but if you leave them for a couple of hours they will become floppy and easier to arrange flat on the paper. Then put on the glass front and secure it so that your subject is held in place. Leave the paper in its frame exposed to light. Over several weeks, you will see the area not covered by your subject fading. When the paper has faded to the level you desire, or for the length of time you have patience for, open the frame and take out the paper. Your subject should be outlined on it. If you want to give your ephemeral art permanence, take a photo of it or scan it straight away. Displayed out of direct sunlight, the original phytotype will gradually fade.

Photographs

John Muir's sketches and photographs by Carleton Watkins and Ansel Adams of Yosemite and other American landscapes are credited with motivating conservation of natural heritage via protected parks. Ansel Adams cited John Muir's writing about American landscapes as inspiration for his photography. Their work was often paired together as depiction of wild nature. The moral and theoretical rationale behind this is being rejected – these were not untouched wildernesses; in fact, they were landscapes whose Indigenous custodians had been recently displaced.

However, the role of photographs in understanding and protecting nature continues. Shifting species distributions are updated through online data aggregators like iNaturalist, where images of wildlife are mapped. Individual basking sharks (*Cetorhinus maximus*) can be identified by their dorsal fins. Inherited characteristics and experience mark their fins with pigmentation, and scars can be caused by parasites or boat injuries. The Shark Trust runs a sightings database for basking sharks in which fin photos are being incorporated in order to track the movements of individuals on their annual migratory journey.

When taking photos of wildlife in and around aquatic environments, the most important principle is to avoid disturbing the wildlife you are looking at. It is too easy to be drawn into the viewfinder or screen and forget the animal you are looking at. Plants aren't disturbed by the act of looking itself, but can be damaged by being trampled. Another important consideration, especially when around water, is to remain aware of your surroundings. Don't step off dry land into water by accident; or, if already in water, stay aware of your position in relation to where you should be and notice if you are drifting. Rivers in full spate, seashores when waves are rough, and being in water that is busy with surfers or boats can be perilous if your attention is too focused on taking a photo without accounting for what is going on around you.

As always, the best camera is the one that you have with you – portability counts for a lot. Luscious close-up photos of birds and wildlife using heavy and expensive zoom lenses are a mission in their own right. The pursuit of perfect images is different from being present in an experience and catching a snippet of it in a photo, and each is valuable in its own right. There are plenty of excellent photography guides that provide technical guidance. Simple but effective advice is to read the instruction manual and understand what your device is capable of, whether it is a phone in a waterproof case or an SLR camera set-up. Some accompanying gadgets are expensive; others are relatively cheap and make a big difference. Polarising filters can transform waterside photos by giving you the ability to select whether the surface of water is dark or if reflections and underwater scenes are revealed. Digital photography also provides a feedback loop that speeds up the process of getting photos you want. Check the photo you have taken on the spot and see if it looks like you hoped it would. If not, try again – or, in the case of moving animals, be ready for the next sighting of them.

Seaweed specimens

During the Victorian era, a zeitgeist of interest in the natural world pervaded society. Pursuing animals and killing them for collection was a manly pursuit. As was botany – since Linnean classification referred to sexual parts of plants seen in flowers and was not considered suitable for women to study. Women were not encouraged to roam far from home by themselves, which also limited the areas in which they could seek wildlife. Seashores offered a hit of wilderness that women could access, since restorative or family seaside breaks were acceptable for women. In seaside towns, in full scrutiny of members of the public, women could walk unchaperoned and reach down into rock pools and retrieve seaweeds to identify; here was nature they were allowed to access. At the time, the complex sexual reproduction cycle of seaweeds was not fully understood. As seaweeds lacked flowers and their sexual parts, their classification and identification was not considered an outrageous pursuit for women to participate in. Women with an interest in the natural world were directed into seaweed collection by social limitations. So women pressed and dried seaweeds, keeping them in collections that were sometimes turned into leather-bound books. Some of these seaweed collections have been preserved in natural history museums. Although the women who made them were not allowed to be members of the Royal Society or the Linnean Society, the engagement they were allowed with natural history has turned out to be a hobby that created not just pretty pressed seaweeds but collections that science still uses today to understand wildlife. DNA analysis of old specimens has helped to advance accurate seaweed taxonomy. A cause of historic fisheries collapse has been identified by analysis of pressed seaweeds, showing cycles in upwelling ocean currents and the influence of human-caused climate change.

Pressing seaweeds uses regular household equipment of a shallow dish, cardboard, old newspaper and scissors, with the addition of blotting paper, watercolour paper, a paintbrush and fine-grade nylon mesh (old tights or insect netting will work). All of these can be reused apart from the watercolour paper that serves as the final mount for the seaweed. Collect sections of seaweed from where they are growing, or use what you find washed up on the shoreline that isn't dried or decomposing.

Small seaweeds are easier to fit onto a sheet of paper. Filamentous seaweeds composed of fine threads look more exciting underwater or mounted on paper than they do when you see them collapsed in a blob on the beach, as water and paper provide support for them to be fanned out. Seaweeds like sea lettuce (*Ulva*) that have a thin texture are more colourful dried and mounted on paper than thicker seaweeds where the pigment is denser. Seaweeds whose structure is more two-dimensional with succulent blades – like pepper dulse (*Osmundea pinnatifida*) – are great for getting to grips with the art of laying seaweeds flat, as they tend to fan out easily and their thickness makes them easier to handle.

At home with your selection of seaweed specimens, put about five centimetres of water in a shallow dish, just enough depth to cover a sheet of watercolour paper with seaweed on it. Use a paintbrush to move the seaweed fronds so they are fanned out, showing their shape in a layout you like. When you are happy with the seaweed's position, slowly lift the paper up so that water drains off and the seaweed remains flat on the paper. It might move as you lift the paper, in which case, while it is still damp and flexible, use a wet paintbrush to nudge the fronds back into place. Put the watercolour paper with the seaweed on it to one side, and layer a sheet of cardboard, newspaper and blotting paper on top of each other. Now put the seaweed on its watercolour paper on top. Seaweed is mucilaginous, and as it

dries will naturally stick to paper. Then add a layer of mesh. The layer of mesh will prevent the top layers of paper from sticking to the seaweed. Add another layer of blotting paper, newspaper and cardboard. Your seaweed is now ready to dry while being kept flat. Repeat the same process for any other seaweed samples you have collected. Stack all of these seaweed sandwiches together in a pile (like a club sandwich) and leave them to dry with weight spread over the surface – books are handy for this. For three or four days in a row – depending on ambient temperatures and how thick and juicy your seaweeds were – go through your seaweed stack and replace the damp blotting paper and newspaper with dry sheets. When it seems that most of the water has been extracted, leave the stack for three weeks to complete the drying and flattening process. Then the specimens are ready to be framed or filed. If you are keeping records of when and where you collected them, write this on the paper. A discreet way to label each sheet if you are framing them is to write in pencil along a margin that will be covered by the frame.

Fossils

Water is adept at revealing ancient life as it washes off silt and dissolves limestone. One of the reasons Lyme Regis in Dorset

Fossilised ammonites can be found whole, and in sections that retain their distinctive ridges

was not just a foundation of palaeontology but continues to yield new fossils is because waves continue to erode soft material at speed and remove denser fossils more slowly. Similarly, rivers and streams where water flow washes soil away have freshly exposed areas that can contain fossils.

Some fossils have the entertainment value of being fossilised poo – an essential but usually ephemeral dimension of animal life preserved by mineralisation. Mary Anning discerned their nature by looking inside and discovering fish remains of bones and scales. Her discovery was ratified by a recognised geologist – and being a man was essential for this role. The early nineteenth century was not ready to recognise women as scientists, and only in 1904 and 1945 respectively did the Linnean Society and the Royal Society admit women as fellows.

Other fossils are brilliantly terrorising;

Megalodon the giant shark and Sarcosuchus the ten-tonne crocodile spring to mind as creatures most people would rather not meet. Many fossils painstakingly unearthed and cleaned look like sculptures. For example, Crinoids look like H.R. Giger – the artist who designed aliens for Ridley Scott – interpreting flowers in biomechanical form. Crinoids are animals that while preserved in fossil records are still found in our seas today.

A walk with a fossil hunter is the best way to pick up the knack for spotting rocks that may contain fossils, learning how to crack them open, and knowing where to check for loose fossils. Natural history museums based near fossil-rich sites and local fossil groups are the places to find an expert guide. Something you might walk past seeing it as a pointed pebble can be quickly identified by an experienced fossil hunter as a belemnite – ancient relatives of squid, their soft bodies decayed but their hard internal skeletons are often found as fossils. Once you've got your eye in, a walk along a shoreline can be enlivened by finding a charcoal dark shark tooth, or an ammonite preserved in glistening iron pyrite.

You should make yourself aware of rules covering land you are on. At Lyme Regis and Charmouth in Dorset, wave action and mudslides are continuously revealing new fossils that can be washed away by the sea. Fossil collection is welcomed, and the fossil centre asks that significant finds – large, rare or held in bedrock – are reported to them. In the setting of a quarry nature reserve, Achanarras Quarry in Caithness is now an SSSI and nature reserve, where fossil collection is limited to collection of loose material only and a maximum of ten specimens per day. An amateur is far more likely to damage a fossil trying to extract it from bedrock than an experienced fossil collector. While fixed to bedrock, fossils are connected to their geological era; taken out of that context, important information on what time a creature lived is separated from the fossil unless the location and details of the fossil are accurately recorded. Significant fossil finds are still shaping how we under-stand life on Earth in past times, and also how life might or might not adapt to the climatic changes that are taking place today.

Perhaps fossils are the prized truffles

of mineral collecting. They have a deep connection to ancient time, and there is also artistry in living forms captured in minerals.

Recording schemes

Currently living creatures and plants are of interest to science as much as past life. Bird conservation has been well informed by decades of the Royal Society for the Protection of Birds (RSPB) collating data generated and collected by bird-loving volunteers who ring and count birds. Every year, the RSPB asks people to pick one hour during their January survey period to join in the Big Garden Birdwatch and record the number of birds they see. Other conservation organisations and research institutes are recognising the value of data collection via the vast network of people who aren't employed as scientists but can contribute as citizen scientists.

After the Big Garden Birdwatch, the next seasonal monitoring event is SpawnWatch. In Japan, spring unfolds with *hanami* (flower-watching), usually cherry blossom and sometimes plum blossom. From south to north, a frontier of unfurling blossom advances and is reported on the national news so that people can enjoy the spectacle when it arrives in their area. Across Britain and Ireland, a similarly advancing frontier is frogs and toads spawning in response to weather. In coastal areas of

the South West you can find frogspawn in January; in the Scottish Highlands it can appear in April. The Freshwater Habitats Trust is one of the organisations that welcome people sending in reports of spawn for their PondNet Spawn Survey.

Throughout the year, the Natural History Museum invites people to send in reports of seaweed sightings. This reporting scheme started in 2009 and is being used to focus on the issues of rising sea temperature, arrival and spread of new-to-Britain species of seaweed, and ocean acidification. Equipped with their seaweed identification guide and following the survey steps, you'll leave your survey site after a briny good time connecting with a five-metre stretch of beach. Without realising it, you will also have been guided into some seaweed identification and have learnt how shifting species distributions reflect broader environmental changes.

Some of our shark species are viviparous and give birth to live sharks. Others lay egg cases – leathery capsules – from which young sharks emerge later. Once juvenile sharks, rays and skates have hatched, their egg cases wash up on shore and often catch in seaweed along the tideline. The Shark Trust has an identification guide to help you distinguish the twirled tendrils and five-centimetre

< Shark and ray egg cases (*see page 178*)

length of dainty small-spotted catshark (*Scyliorhinus canicula*) egg cases from twenty-eight-centimetre-long white skate (*Rostroraja alba*) egg cases and everything in between. The Shark Trust are also keen to receive reports of egg cases spotted by the growing number of divers, snorkellers, swimmers, sea kayakers, stand-up paddleboarders and seashore dabblers who spend time in water and see shark egg cases where they have been laid attached to seaweeds. Reporting finds of these egg cases helps to map distributions of sharks, rays and skates around our coasts. In particular, it helps to identify nursery habitats that are critical conservation targets for ensuring the survival of these species. According to the International Union for Conservation of Nature (IUCN) Shark Specialist Group, thirty-one per cent of sharks and rays are threatened.

Jellywatch collates reports of jellyfish sightings around the world. Jellyfish blooms have economic impacts and associated social consequences on fisheries and tourism. Some of our charismatic visitors such as the leatherback turtle (*Dermochelys coriacea*) are increasingly sighted in the Atlantic and around our shores, particularly in August, the peak of our jellyfish season. In oceanic food webs, jellyfish are influential, and their distributions and numbers are thought to be changing. Jellywatch is seeking to understand possible drivers of the expansion of jellyfish blooms, examine their effects, and identify current and future consequences of them. Answers to all of these questions get closer with every jellyfish sighting added to their database.

PlantTracker was a collaboration to monitor the spread of plants considered a cause of concern because they had spread rapidly since being introduced, and they displaced other plants. Many of the species of concern were aquatic or grow in damp or water-adjacent habitats, benefiting from taking advantage of water as a fluid transport system. They included giant hogweed (*Heracleum mantegazzianum*), Himalayan balsam (*Impatiens glandulifera*), American skunk-cabbage (*Lysichiton americanus*), monkey flower (*Mimulus guttatus*), curly waterweed (*Lagarosiphon major*), water fern (*Azolla filiculoides*), water primrose (*Ludwigia*), New Zealand pigmyweed (*Crassula helmsii*), parrot's feather (*Myriophyllum aquaticum*), orange balsam (*Impatiens capensis*), giant rhubarb (*Gunnera tinctoria*) and floating pennywort (*Hydrocotyle ranunculoides*). Combining crowd-sourced data with the convenience of a phone app that mapped photo-verified reports resulted in over 25,000 records for scientific analysis and guiding interventions. Smartphones are an innovation that have been embraced by many ecological and conservation groups and initiatives because their ability to take a photo and tag it with GPS data creates high-quality verified records.

Capturing memories is both a construction of our individual selves and a contribution to collective memory and legacy. Current and future generations suffer from the loss of environmental knowledge held by the community. Because our environment is rapidly changing between generations, we are downgrading our perceptions of what normal environments and wild places are like. One of the best ways to combat these shifting baseline perceptions which disguise what should be alarming warnings and aid lack of action is to consciously record what we see.

There is a dilemma presented by our presence in wild places. We come to enjoy what they are, but our presence risks disturbing and changing them. Knowing more about the creatures and plants that inhabit them informs us about how to behave better. We can both enjoy wildlife and be good custodians of wildlife and the space it inhabits; nature is not a museum.

Leatherback turtle
(*Dermochelys coriacea*)
and moon jellyfish
(*Aurelia aurita*) >

6

Wildlife

Ambergris

Ocean alchemy

While alchemists strive to create gold, sperm whales (*Physeter macrocephalus*) and pygmy sperm whales (*Kogia breviceps*) dump floating gold in oceans. Sperm whales range across the equator up to the edges of pack ice in the far north and south. Ocean depths are their hunting grounds; squid constitute a large part of their diet. Whales usually dispose of squid's hard beaks by vomiting. Sometimes they don't, and squid beaks travel through the whales' four stomachs into their intestines, where they are lubricated with ambrein and saturated with faeces, and eventually become a lump of ambergris. In fortunate instances, the whale will pass this out when

< Salmon (*Salmo salar*) swimming upstream
  in their seasonal migration

it defecates. Otherwise, it remains inside the whale, slowly growing until it causes a fatal rupture to the whale's intestines. Unleashed along with poo or when scavengers have eaten the rest of a whale's carcass, ambergris floats on the sea and occasionally washes up on our shores.

Despite having either gross or grisly origins, ambergris has been valuable to people for at least the past 1,000 years. Sanskrit records from AD 700 refer to ambergris as a perfume, it is an ingredient in a seventeenth-century ice cream recipe and it is used in contemporary perfumes. Aged by floating in seawater under hot sun and chill winds for years, ambergris pales from black to grey, and transforms from faecal-smelling into an earthy aroma.

Formed infrequently, taking years to

Sperm whale (*Physeter macrocephalus*) and ambergris

mature and being found when it washes up on beaches makes ambergris a rare and valuable ingredient, hence its alternative name of 'floating gold' and headlines quoting six-figure sums when people find ambergris on beaches. In some countries, possessing or trading in ambergris is restricted, for instance under the Marine Mammal Protection Act and Endangered Species Act in the USA, or Australia's Environmental Protection and Biodiversity Conservation Act. But the Convention on International Trade in Endangered Species treats ambergris as an excretion and therefore does not limit its trade. In Britain it is legal to collect and sell ambergris found on a beach.

Ambergris floats in seawater, which differentiates it from the stones that it resembles. A hot needle pushed into ambergris will melt what it touches, and when the needle is pulled out a tacky

residue will cling to it. Lumps of solidified sewer grease or palm oil also wash up on beaches and respond similarly to a hot needle, but in contrast ambergris has a fragrance ranging from marine to earthy.

Where to look for sperm whales and ambergris

> Bay of Biscay – take a ferry from the south coast of England to northern Spain and watch out for sperm whales in deep waters of the southern and western areas of the bay.
> Anglesey, Wales – pygmy sperm whales are rarely seen, but one was sighted in 2014. In 2015 ambergris worth £11,000 was found on a beach in the north.
> Hengistbury Head, Dorset, England – ambergris was found here in 2012, but the surrounding sea is too shallow for sperm whales.

> North-west Scotland – sperm whales have been seen off the coast in summer, and winter sightings are now also being reported.

Atlantic puffins

Life divided between earth, sea and air
Summer is when we see puffins as they give up their oceanic nomad state to rear chicks on land. Every Atlantic puffin (*Fratercula arctica*) has an individual winter voyage that follows a similar pattern each year. They wander alone on their personal itineraries. One puffin might winter in Iceland and Majorca, while another flies from Ireland to the Greenland Sea. These solitary winter puffins sit on the sea and dive to eat fish, wearing grey-faced winter plumage. Their feathers are waterproofed and they can drink seawater. Atlantic puffins don't need land, until the weather warms and they return to shore like heralds of the summer.

Puffins begin summer by congregating in rafts of mating birds. Despite the multitude of options, puffins tend to be monogamous and mate with their partner from previous years. Although fidelity is high, these birds still make an effort with their appearance. Carotenoids, fat soluble pigments absorbed from fish, are concentrated in their bill, legs and feet, which turns them bright orange. This is a signal indicating prowess at catching fish that helps to make them attractive to the opposite sex. Orange is so alluring to puffins that hunters used to paint rocks orange as bait and wait nearby in orange clothes.

Each pair of puffins has just one egg per year. Nesting underground offers protection for both egg and young chick from raiding gulls. Puffins cut soil with their sharp beaks and shovel it with their feet to excavate burrows. They dig with intention and design in mind. In addition to the nest area, where the egg is laid and the chick matures, burrows have a toilet area.

In breeding season when puffins have chicks to feed they spend about seven hours underwater each day. They dive between 600 and 1,150 times a day to catch fish. Most of their dives are around fifteen metres deep. But they can go down as far as sixty-seven metres and spend two minutes flying underwater to hunt. Their wings are relatively short, which reduces drag when using them underwater. But this means puffins can't glide on air currents and they must beat their wings frantically to stay aloft in the air.

Atlantic puffins are widespread – ranging across the Atlantic – and there are about 12 to 14 million of them. But their numbers are decreasing, and unless the causes of their decline ease, puffins are

Atlantic puffins (*Fratercula arctica*)

likely to become an endangered species. Hunting in Iceland and the Faeroe Islands, where puffins are eaten, contributes to their troubles. Visits by people to nesting areas can decrease their breeding success by forty per cent. But for these birds built to fly underwater, not to travel long distances by air, climate-change-induced shifts in the distribution and abundance of herring and sand eels are causing some puffin colonies to collapse. Their habitual summer colonies are becoming too far from hunting grounds.

Where to see Atlantic puffins

Puffins congregate on land from spring to August. Watching them from a distance with binoculars, in a boat or in the water causes less disturbance to parents delivering food to a hungry chick than walking around their colonies by foot.

> Skomer Marine Reserve, Pembrokeshire, Wales
> Rathlin Island, County Antrim, Northern Ireland
> Bempton Cliffs, Yorkshire, England
> Bullers of Buchan, Aberdeenshire, Scotland
> Lunga, Treshnish Isles, Argyll and Bute, Scotland

Barrel jellyfish

Gentle giants

In 2015, photos of a sea swimmer called Jane smiling with delight while underwater next to a huge jellyfish spread around the world from a local south-west England newspaper to British nationals and the *Daily Pakistan* newspaper. That jellyfish was living up to its name of dustbin-lid jellyfish or barrel jellyfish (*Rhizostoma pulmo*). They can have domes that are just shy of a metre wide, and they can weigh up to thirty-five kilograms. Perhaps what was most captivating about the photo was not the size of the jellyfish but the clear delight on Jane's face, when fear of jellyfish is common enough to have a name – cnidarophobia. If you are one of those swimmers who is horrified by jellyfish, barrel jellyfish might be the ones to help you.

Despite being joined by barrel jellyfish on sea swims every summer, I haven't been stung by one. Firstly, they are so big that even when the water is a bit murky you can see them in front of you. Their great size isn't a cue for terror, rather a feature that makes it easier to avoid them. Secondly, the only part that stings is the tentacles and they just have eight relatively stumpy tentacles, which makes it easy to avoid the stinging parts, particularly when the non-stinging dome is nearly half of the mass of the jellyfish. Despite their size, barrel jellyfish do not have a sting that is powerful enough to be notable for a human. I did once accidentally get a handful of barrel jellyfish tentacles, but didn't feel a sting. More delicate skin, for example on your face or skin exposed by skinny-dipping, might feel a sting from them. So avoid touching them, but perhaps more so that they can keep their sting for their prey and not waste it on you.

When the barrel jellyfish come to visit, my friend Keith is up for 'mission jellyfish'. He loves to capture them in photos. Their

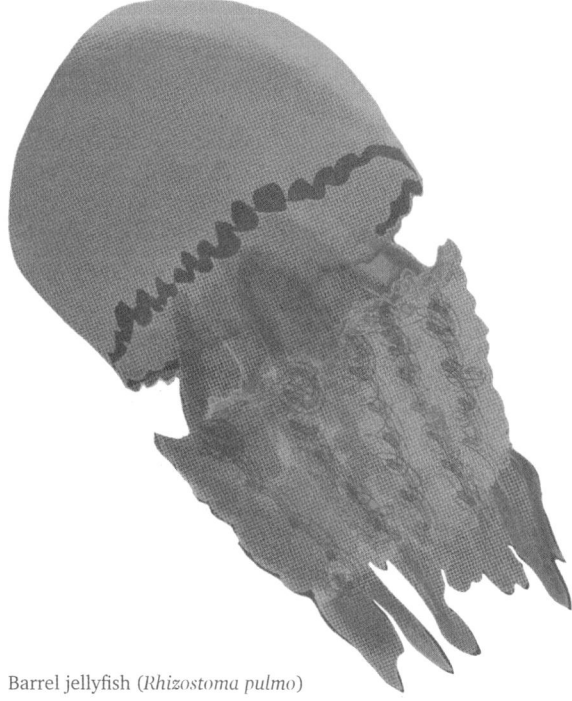

Barrel jellyfish (*Rhizostoma pulmo*)

bodies are translucent with a violet line around the edge of their dome, particularly beautiful if you catch sunlight shining through them. Perhaps they are the optimum subject for photos of wildlife seen while swimming, as they are not disturbed by people being near them and they are so big that it is easy to get a camera to focus on them.

Barrel jellyfish appearing offers an enticement to visit our shores for the animal that loves to eat them: leatherback turtles. If they are good enough for the world's biggest turtle to eat, would you eat them? They are full of proteins that have high antioxidant activity. Barrel jellyfish are considered to be edible for people too, but preparing them is a long process of dehydration and brining in which salt concentrations need to be calculated correctly. I feel like I enjoy swimming with these creatures more than I want them on my plate.

Where to see barrel jellyfish
> Sandbanks beach, Poole, Dorset, England
> Loch Long, Argyll and Bute, Scotland
> Greatstone beach, Kent, England
> Summer Isles, Highland, Scotland
> Freshwater East, Pembrokeshire, Wales

Basking sharks
Sharks without bite

Appearing in the manner of the Loch Ness monster, emergent tips of dorsal fin, tail fin and snout are often visible on the surface from land or boat when basking sharks (*Cetorhinus maximus*) are in the sea. Underwater, you can see up to twelve metres of body length between snout and tail fin. Reaching that size on a diet of microscopic creatures – plankton – takes decades. However, basking sharks have mastered the art of passive eating. By slowly swimming with their mouths gaping open, they can filter 1,800 tonnes of water an hour while sieving out the plankton. They are certainly sharks that do not consider people to be edible.

Basking sharks inhabit temperate waters in both the northern and southern hemispheres. We see them demonstrating how they got their name – basking at the sea's surface. But shark tagging reveals what they do when we don't see them. One basking shark tagged off the Isle of Man travelled to Newfoundland. On her 9,589-kilometre journey she also swam into deeper waters, as deep as 1.264 kilometres.

Being able to dive deep was a factor in the decline of basking-shark populations. Oil in their livers helps them to regulate buoyancy. Oil-rich shark livers were consumed by the oil-lamp trade until kerosene replaced oil for lighting. Basking-

Basking shark (*Cetorhinus maximus*)

shark livers are still used as a source of squalene. In skincare, hydrogenated squalene is called squalane and valued for its emollient characteristics. Fortunately, vegetable sources of squalane are gaining currency. So, before buying a product containing squalane, like sun cream or lipstick, check with the manufacturer that it is vegetable-derived. Animal-derived squalene remains in use as an adjuvant (an additive that enhances immune response) in vaccines. Pharmaceutical use of squalene is smaller than cosmetic use, but is a concern because it is easier to substitute ingredients in cosmetic formulations than in medical usage. For plant-based alternatives to be used in applications like vaccines, which have global health significance, their efficacy as adjuvants must be proven.

Where to see basking sharks

On calm days when the sea is smooth it is easier to spot their noses and fin tips at the surface, and on sunny days there is better visibility underwater. Off headlands or at the mouth of a bay are likely places for warm water to meet cool water, which is a condition plankton concentrates in – an attractive location for a basking shark to be. Basking sharks are around the British Isles from April to October, but peak season for spotting them is mid-May to July. Early on, they will be at southerly locations; as summer progresses, they head north up to Scotland. Remember that they are a legally protected species. You can be prosecuted if you breach the Basking Shark Code of Conduct under Schedule 5 of the Wildlife and Countryside Act 1981. Moreover, breaking the code that has been written to protect them could cause them distress or have negative impacts by disrupting their movements and feeding. If you are in the water, stay at the surface rather than diving down to look at them. Don't touch basking sharks and keep at least four metres between you and any basking shark (there can be more than one present). Groups of swimmers should stay together, and there should not be more than four people within 100 metres of a basking shark.

> Newquay, Cornwall, England
> New Quay Bay, Ceredigion, Wales
> Blasket Islands, Dingle Peninsula, County Kerry, Ireland
> Bradda Head on 'Basking shark capital of the world' the Isle of Man
> Oban, Argyll and Bute, Scotland

Beadlet anemones
Personality and immortality

At low tide, they look like glistening red and green sweeties stuck on rocks. When the tide rises, tentacles emerge and beadlet anemones (*Actinia equina*) release their aggressive impulses. Bright blue dots underneath their tentacles, as if they are wearing a necklace of beads, are where their stinging powers reside for claiming territory. When touched by an anemone that is genetically different, beadlet anemones inflate these blue dots, called acrorhagi, revealing them to be specialised tentacles. Acrorhagi are exclusively used for fighting other anemones. The standard tentacles surrounding their mouth are red or green – the same colour as their central column – and have stinging cells that are used for immobilising prey before eating it. They eat anything they can catch and cram into their mouths, from small fish and shrimps to crabs.

Watching beadlet anemones fight, scientists have observed that winning is not just determined by size and strength; personality has influence. Startling beadlet anemones with a jet of water makes them hide by retracting their tentacles. Beadlet anemones that have won fights quickly recover from being startled, while fight losers remain withdrawn for longer. Startling anemones with water is a crude method, but it reveals that beadlet anemones aren't solely ruled by reflexes – they have an inner world.

With their antisocial tendencies, it is no surprise that beadlet anemones reproduce while maintaining distance. Males release sperm and if it reaches females, eggs are fertilised and develop until being launched as miniature anemones. Male, female and asexual beadlet anemones can also propagate themselves by cloning. Small buds develop inside them, which grow into independent anemones that share identical genetic material with their sole parent.

Beadlet anemones (*Actinia equina*)

Beadlet anemones inhabit intertidal zones, clinging to rocks that are under the sea for at least part of the day but may also be dry during low tides. Living between two worlds is tough. Beadlet anemones have to cope with salinity and currents when immersed in the sea. They also have to survive bright light and temperature extremes while being exposed to air. One trick is producing mucus that coats their soft bodies and protects them from being thoroughly desiccated by air. Although they look immobile, beadlet anemones can move. Like slugs and snails, they are aided by their mucus providing a slippery surface to glide on.

Sea slugs and some species of fish are known to eat beadlet anemones, but if a beadlet anemone isn't eaten and water quality remains decent, there is no reason for it to die. Lost tentacles can be replaced. If their mouth is cut off, they can grow a new one. Cells are replicated without mutation or slowing of pace. As time passes, beadlet anemones do not age.

Where to see beadlet anemones
Keep in mind that rock pools covered by high tide are booby traps for getting skinned shins if you accidentally swim into rocks. So if you visit sea anemones by swimming, rather than rock-pooling, watch where you are going.

> Gouliot Caves, Sark, Channel Islands
> West Runton beach, Norfolk, England
> Barmouth beach, Gwynedd, Wales
> Flaggy Shore, County Clare, Ireland
> East Sands, St Andrews, Fife, Scotland

Brown trout

One fish, many forms

If one creature can show us that genetics doesn't entirely explain the differences in form that life takes, it is brown trout (*Salmo trutto*). Little freshwater-dwelling brown trout, sea-faring trout and giant ferox trout are all the same species.

Golden–brown backs, yellow bellies, and red and black spots make brown trout far more colourful than their name suggests. They inhabit well-oxygenated, cool and fast-flowing freshwater. They catch their diet of small fish and insects by ambush. When it is time to spawn, they mingle with fellow brown trout, and also with visiting sea trout.

Sea trout spend the first year or two of their life being indistinguishable from brown trout. But then they change. Outwardly they turn silver; inwardly they become able to live in saltwater. Gathering together, these salt-adapted trout migrate to the sea. Finding richer sources of food there allows them to grow bigger than brown trout. Sea trout return to their natal river to spawn. While closely related salmon tend to perish after spawning, most sea trout retain enough vitality after spawning to travel back to the sea.

Ferox trout take us to an ancient era just after the Ice Age, when the glacial lakes they live in were formed. Despite living in nutrient-poor cold lakes, they become far larger than brown trout. Juvenile ferox trout are more aggressive and control better feeding territories than juvenile brown trout. Although ferox trout grow more quickly than brown trout, ferox trout mature later. So when they do spawn, ferox trout are bigger. They eat sizeable fish, not just little fish and insects. Arctic charr are their main prey, but ferox trout adapt to eat other fish, even brown trout if Arctic charr are absent. Ferox trout don't just ambush prey – they will also pursue their desired food. They can even be considered the apex predator in their lakes.

Most ferox trout are thought to be the same species as brown trout. But in Scotland's Loch Awe and Ireland's Lough Melvin, ferox trout are genetically distinct from the brown trout inhabiting the same water. These ferox trout are thought to have ancient lineage dating back to early colonisation of their lakes. Different spawning and feeding habits have kept them from merging with more recently arrived trout.

When and where to see brown trout

Trout fishing season has some regional variation, but runs from March until the first week of October. During autumn and winter swims in trouty rivers, you are

Brown trout (*Salmo trutto*) >

less likely to come into conflict with trout fishermen who sometimes pay a lot for their fishing licences. But at this time there is extra responsibility to avoid trampling gravel where trout have left their eggs to hatch. One of the best ways to see river fish also causes them the least disturbance: drifting downstream with the current. Because you aren't making movements that startle them, fish stay within closer view. River swims are good for brown trout year round, and for sea trout from June until October.

> Ferox trout and brown trout: Loch Rannoch, Perth and Kinross, Scotland
> River Wharfe, Appletreewick, Yorkshire, England
> River Arun, Sussex, England
> Monnow Gate and Bridge, Afon Mynwy, Monmouthshire, Wales
> River Till, Northumberland, England

By-the-wind sailors

Ocean voyagers

Wintery beach walks with my dog yield an assortment of flotsam and jetsam on the tideline: oranges, nurdles and sometimes what looks like discarded toenails. Closer inspection finds concentric rings and a triangular fin on the surface, not manky toenails after all. It takes a storm to reveal the iridescent creatures that create these vessels with which they sail oceans and seas.

Needing neither rocks to cling to nor shelter amongst seaweed, reefs or wrecks, by-the-wind sailors (*Velalla velalla*) thrive in the freedom of open water. Being carnivores that are only two to eight centimetres long, their prey is tiny crustaceans one to two millimetres big, fish eggs and larvae. While by-the-wind sailors, along with other members of Cnidaria like jellyfish, are equipped with tentacles, their sting is adapted to small prey and their

tentacles are short. So, although they float in fleets, they are an armada without sting for people. They also host photosynthetic microalgae that provide them with additional nutrition.

Having bodies shaped perfectly for floating, by-the-wind sailors spend their entire adult life on the surface of the sea, and are only underwater during their immature larval stage, when they look like tiny jellyfish. A triangular fin runs diagonally across the upper surface of their oval bodies. Most by-the-wind sailors are lefties: their fin runs from the upper left down to the lower right of their body. Some are right-orientated, with their fin running in the opposite direction – from right down to left. Catching wind with their fins directs them to travel downwind or at a slight angle to the wind. With propulsion controlled by wind, they cannot avoid being blown ashore. As seventy-one per cent of the Earth's surface is covered by sea, there is enough open water for this marine species to persist even when thousands of individuals are stranded on beaches.

Like many surface-floating sea creatures, they are tinted violet–blue, a colour speculated to offer the advantages of camouflage in salty water and protection from exposure to the sun. This colour is

< By-the-wind sailors (*Velalla velalla*)

responsive to temperature and salinity; washed ashore, by-the-wind sailors fade, and as they dry out, toenail-like husks are all that remain.

When and where to see by-the-wind sailors
After strong southwesterly winds have blown in across the Atlantic Ocean, by-the-wind sailors can be seen in our shallow coastal waters and on shore. To get a good look at their iridescent colours on a beach requires crawling on hands and knees to get down to their level. In the sea, floating at eye level, it is easier to admire them. By-the-wind sailors are camouflaged in colour, but since they travel en masse and their shape breaks up the surface of the water, they are easy enough to spot.

> Inch beach, County Kerry, Ireland
> Sandwood Bay, Sutherland, Scotland
> Porth Neigwl (Hell's Mouth), Llŷn Peninsula, Wales
> Lulworth Cove, Dorset, England
> Westward Ho! beach, Devon, England

Caddisflies
#vanlife as an insect
Building their own portable case to inhabit while roaming means caddisflies have had a mobile lifestyle since before social media gave currency to the idea that life on the road was good living. For caddisflies, their

Caddisflies (*Lepidostoma basale*)

mobile home phase of life is a time of growth within shelter, before they take to air as winged adults.

Of about 200 species of caddisfly living in Ireland and Britain, all but one spend their larval stage submerged in water; the exception lives on forest floors. About half of these species are cased caddis that make portable shelters. Uncased caddisfly make fixed shelters or nets to surround themselves, or roam as voracious predators unrestricted by casing. Identifying cased caddisfly larvae is speeded up by looking at what they use to make their case – each species is quite discerning and only uses specific kinds of material.

Lepidostoma basale use sediment grains to build their cases by sticking them together with a silk lining they spit out as silk thread from glands by their mouths. Not only does

their silk manage the feat of acting as a glue underwater, it creates structures that can withstand force by being able to expand and contract. They used to build just using their silk thread combined with sand and tiny pieces of gravel, sediment naturally found in rivers and streams where they live. But now they also use microplastics. Small plastic spheres used as exfoliating agents in cosmetics have been found glued into *Lepidostoma basale* cases. With the recent ban on using plastic in these products, that may reduce over time. However, most of the plastic found in their cases is secondary fibres, films and fragments created when larger pieces of plastic degrade. A high proportion of this is polypropylene, which is used in single-use plastics. As these insects build their cases with materials around them, they work as a bioindicator of pollution. They are an indicator that we need to #take3forthesea and pick up plastic upstream from beaches too.

Living in a sturdy case provides caddisflies with protection from predators. *Lepidostoma basale* eats some insects, wood debris and biofilms comprised of microorganisms that are stuck together. Eventually, caddis larvae grow big enough to pupate and turn into winged adult insects. To reach the air, they bite their way out of their case and float up to the surface. The transition between case and air is a vulnerable moment and many become fish food at this point. Once they are airborne, they join the host of winged insects found around waterways. It is in their juvenile form inside their cases when they are most distinctive.

Where to see caddisflies

In clear-running streams and rivers, take advantage of the magnifying effect of goggles and water to look at the streambed or river bottom. Watch for insect eyes peeking out of a little tube of sand and gravel.

> River Feshie, Highland, Scotland
> Kielder Burn, Northumberland, England
> River Len, Kent, England
> West Beck, Yorkshire, England
> River Clwyd, Denbighshire, Wales

Cuckoo wrasse

Sometimes sneaky, newly valuable

On the list of changes to make in order to become male, colour doesn't spring to mind for most people. But for cuckoo wrasse (*Labrus mixtus*), being bright blue and orange is essential for attracting females who prefer intensely coloured males. Every male cuckoo wrasse has transitioned from being female, since every cuckoo wrasse is born female. When wrasse reach twenty-six centimetres long, if there is a dearth of males in the area, female cuckoo wrasse become male.

Between five and twenty per cent of male cuckoo wrasse are indistinguishable from pink–orange-coloured females in outward appearance and use a different tactic to breed. Instead of attracting mates, males clad in female colours sneak to nests built by standard blue–orange males where eggs have been laid and try to fertilise them. Ecologists call this sneaky strategy kleptogyny, and point out that it shows how subordinate males can be successful.

It is commonly accepted that cuckoo wrasse are not of commercial interest because they are not edible. However, UK fish licensing authorities have had to respond to a new trade pressure that peaked with live wrasse being sold for £17.50 each, and a tonne being worth £50,000. Cuckoo wrasse are valuable alive because of the work they perform alongside

Cuckoo wrasse (*Labrus mixtus*)

other species of wrasse in the salmon-farming industry.

Sea lice are parasites that naturally live on salmon. When salmon are forced into high densities in fish-farming pens, sea lice reduce the health and growth of salmon to an extent that is noticed in profit loss. Chemicals are used to control sea lice, but periodically sea lice become resistant to the chemicals used. Wrasse put into salmon pens work as cleaner fish for the biocontrol of sea lice. It seems like a brilliant, environmentally friendly solution. Except it relies on wild-caught wrasse. In just eighteen weeks in 2015, 57,000 wrasse were taken from the sea between Weymouth and Lulworth in Dorset and sent to salmon farms. Many wild-caught wrasse die during storage and transportation, and it is thought that cuckoo wrasse have a particularly low survival rate.

It is very difficult to calculate what might be a sustainable level of taking wild wrasse from the sea; since they aren't eaten and haven't previously been considered valuable, there aren't records of their populations or the impact of fishing on them. Our fishing licensing authorities have started to restrict wrasse collection – for example, Cornwall's live wrasse fishing byelaw issues only five permits a year and cuckoo wrasse are excluded. In 2019 the island of Sark became the first place in the UK to ban live wrasse export.

Where to see cuckoo wrasse

These are the kind of fish you'll see on a leisurely swim or snorkel along rocky shores where looking at the scenery is more of a focus than the distance swum. They are inshore fish, not open ocean dwellers, and stay near the underwater features that provide them with shelter and food.

> Booby's Bay, Cornwall, England
> Dunseverick harbour, County Antrim, Ireland
> Seil island, Argyll and Bute, Scotland
> Marloes Sands, Pembrokeshire, Wales
> Valentia island, County Kerry, Ireland

Common cuttlefish (*Sepia officinalis*)

Cuttlefish

Inkfish

Vintage-style photo apps wouldn't look the way they do without common cuttlefish (*Sepia officinalis*). Although apps and software swim in seas of code while cuttlefish inhabit salty water, they are connected by the use of cuttlefish ink in the 1880s to develop photos. Prints developed using cuttlefish-derived sepia toner remained monochromatic, but stark white highlights were softened with a wash of red–brown tint. By converting silver bromide into more stable silver sulphide, sepia toner from cuttlefish ink also enhanced photograph longevity. Long before photography was invented, sepia ink was used by the Ancient Greeks and Romans for writing. Cuttlefish aren't the only cephalopods to produce ink coloured by melanin, the same group of pigments that human skin and hair is coloured by. Octopus ink is black, squid ink blueish–black and cuttlefish ink is reddish–brown. Let's call cuttlefish the redheads of the cephalopod realm.

Cuttlefish ink isn't confined to literary and photographic arts; it is also the key ingredient for *riso nero*, literally 'black rice', a dish particularly associated with Italy and Croatia, countries that abut the cuttlefish-rich Adriatic Sea. Cuttlefish ink adds depth and briny flavour as well as colour. In most scenarios, apart from being used by humans in cooking and visual arts, ink is a functional tool for cuttlefish. When threatened by predators, they squirt out a smokescreen decoy of ink, which gives them opportunity to escape.

Taking less than a second to change their appearance, cuttlefish are expert chameleons. Like other animals, they can change their position and seek cover in an attempt to blend in with their background. Just as rapidly, cuttlefish change not only their colour but also the patterns on their skin by using muscular contraction to change the size of chromatophores, pigment-filled cells. Another tool in their repertoire of optical illusions is the ability to change surface texture between smooth and rough – a bit like extreme goosebumps consciously controlled.

Where to see cuttlefish

Cuttlefish are masters of disguise, but you can still see them. If you sea-swim, the odds are high that at some point you'll see flat, pointed-oval-shaped cuttlefish bones washed ashore. Usually these cuttlebones arrive en masse as cuttlefish die after breeding and their bones, technically internalised shells, float ashore. Cuttlebones are made of aragonite, a lattice of calcium carbonate, which provides cuttlefish with buoyancy control as they fill gaps within its structure with gas to rise and liquid to sink. If you own a budgerigar, you might have put a cuttlebone in their cage for them to peck on and ensure that they get enough calcium in their diet. Another common way to sight cuttlefish is via their eggs. Looking distinctively like a bunch of grapes washed up on the tideline or attached to seaweed underwater, the eggs are black and found in clusters. Cuttlefish are harder to spot alive and camouflaged in the sea, but perhaps more of a delight, especially if you see them magically change colour.

> Weasel Loch, near Edinburgh, Scotland
> Staxigoe, Highland, Scotland
> Porlock Bay, Somerset, England
> Botany Bay, Kent, England
> Babbacombe Bay, Devon, England

Daubenton's bats

Water bats and batty swims

Being buzzed by hordes of bats might sound like an excerpt from a vampiric horror movie. Rather, it was an unexpected display of speed and delicacy skimming over our river swim. While many swims get named by location, that one was named by experience: Amanda still calls it 'The Batty Swim'. While we drifted leisurely downstream with the current, bats whizzed past and overhead like dark shooting stars against a dusky lavender sky.

Daubenton's bats (*Myotis daubentonii*) don't swim, yet they cannot live without wild water. At dusk they emerge from their daytime roosts, which are often in tunnels or under bridges, to fly over rivers and lakes hunting for caddisflies, mayflies and small flies like chironomid midges. It's no coincidence that a fine trout-fishing spot is often a good location for spotting Daubenton's bats; both fish and bat share a taste for insects that lurk at the surface of freshwater.

It might seem funny to describe a creature that weighs about as much as an AAA battery as the owner of large feet, but Daubenton's bats do have large feet relative to their size. Their feet are not clowns' feet; they are predators' feet functioning as a tool in their hunt, since Daubenton's bats catch their prey while in flight, sometimes snatching insects from the water's surface with their feet.

All our native bats are insect-eating and not blood-sucking. Blood-sucking bats mostly inhabit films. You would need to go to Central and South America to

Daubenton's bat (*Myotis daubentonii*) >

meet vampire bats (*Desmodontinae*), and even there reports of them feeding on people are rare. Our native bats will avoid accidentally flying into you, since they are aiming for food. They use echolocation to detect objects, so even in the sparse light of dusk or complete dark they will head for their prey and not people. Essentially, they fly around shouting, and decipher the echoes returning from their racket to know what is ahead of them. This biological form of sonar is conducted at such a high frequency that adults can't hear it, so batty swims are also peaceful swims.

Bats are legally protected species in Europe and it is an offence to disturb their roosting places or to capture them. An evening swim in a river or lake is the liminal space in which swimmers can enjoy prime views of these creatures that inhabit the air over freshwater.

Where to see Daubenton's bats
Skimming over water in the twilight hours.

In early autumn they have a frenzy of activity, as they mate in September and fatten up in October, before hibernating in November underground in caves and mines. Daubenton's bats re-emerge from winter roosts in spring. Remember to check out swim entry and exit points in daylight before using them for lowlight or dark swims.

> River Wharfe, Yorkshire, England (but not at The Strid, which is known to be a perilous swimming spot)
> River Test, Hampshire, England
> River Coln, Gloucestershire, England
> Upper Glendalough, County Wicklow, Ireland
> River Annan, Dumfries and Galloway, Scotland

Dippers
Singing swimmer

If the person who named Britain's aquatic songbird had seen it underwater, they would probably have called it mercury bird or silver swimmer. When immersed, these birds wear a fluid cloak of silver bubbles clinging to their feathers, making their underwater walks look mercurial. Instead, dippers (*Cinclus cinclus*) were named for their terrestrial tail-wagging and dipping displays. Seeing dippers on land or in the air is only part of their story.

Dippers are birds augmented with attributes that enable them to walk on riverbeds under the pummelling force of upland waters rushing down inclines. These are not birds of languid lowland rivers. Keeping themselves greasier than the average bird requires an extra-large preen gland. Dippers' well-oiled feathers trap bubbles underwater. When their brown and white markings are silvered by bubbles, dippers are better concealed from their prey. Apart from an advantage of disguise while hunting underwater, their greasy feathers shed water with ease when they leave their underwater domain. With fifty per cent more feathers than a blackbird, they are also well insulated for time spent in cool water.

Walking underwater is made easier by other adaptations dippers possess. Their claws have strong grip and they have long toes, which makes it easier for them to cling to riverbeds and resist the force of river currents. Underwater, their nose is sealed with a flap, preventing water from flowing into their lungs. At the level of molecular biology, dippers are a bit special: having high levels of haemoglobin in their blood means that they can store more oxygen than most birds, which is helpful for staying submerged.

Hunting underwater is made easier by eye muscles that can control lens shape to overcome visual distortion caused by light refracting below the surface. Dippers eat

Dippers (*Cinclus cinclus*) >

insects that live in rivers – such as stonefly, mayfly and caddisfly larvae. Most of their diet consists of prey that requires fast-flowing, well-oxygenated water. This kind of habitat is found in upland rivers and some streams and rivers in south-west England. Having spent my childhood in the south-east and East Anglia, it wasn't until I went to Wales as an adult that I had my first dipper sighting. After initially thinking I'd seen a bird drown in a rushing torrent, I realised it was deliberately walking on the riverbed.

Dippers are a living barometer of water quality. As you might expect, they did not thrive in rivers that ran with colliery waste. The decline of coal mining allowed rivers to revive even in the heart of mining conurbations; now you might even spot a dipper in Sheffield or Teeside. Paradoxically, rural habitats are becoming less hospitable for dippers. Industrial poultry farms release nutrients from chicken waste into waterways, which fuels rapid algal growth that depletes oxygen

levels in water. Dippers are influenced by this indirectly as their prey declines and rivers cannot provide sufficient food for as many of them as they used to.

Where to see dippers
> Glen Affric, Inverness-shire, Scotland
> Clydach Ironworks, Monmouthshire, Wales
> Glendalough, County Wicklow, Ireland
> River Ribble, Lancashire, England
> Upper River Tees, Durham, England

Eider ducks
Not your average duck

Chunky ducks, sea ducks, friends with Vikings; winter brings more of these quirky ducks to our coastal waters. Keeping warm Viking-style uses eider ducks (*Somateria mollissima*). Putting aside thoughts of duck slaughter, instead turn to thinking about how resourceful Vikings were in making use of natural materials. And be relieved that in Iceland and Norway, people selling eiderdown find that the easiest way to

collect it is just picking it up from eider duck nests at the end of the nesting season.

Canny harvesters wanting to reap this harvest year after year take care to look after wild eider ducks. They build houses and nests for them. Cats and young children are kept away from nests, and people guard eider ducks from fox and mink at night. Around human settlements, eider ducks congregate in far bigger colonies than when they nest in remote places. These ducks have learnt that a mutually beneficial relationship is possible between people and birds. For at least 1,000 years since Vikings ruled northern seas, people and wild eider ducks have had a mutually sustaining relationship. There are clear financial motivations for the labour of custodianship over wild eider ducks, but also cultural meaning. One of the reasons the Vega archipelago in Norway is a UNESCO World Heritage Site is for the relationship that still exists between people and eider ducks there.

Left to their own devices, eider ducks build their nests from grass and twigs. Humans, both lacking the dexterity required for nest-building with twigs and having an ulterior motive, build eider duck nests with seaweed. It is much easier to clean the down collected from nests made with seaweed. It takes down from about sixty nests to make one eiderdown quilt.

< Eider ducks (*Somateria mollissima*)

A good motivation for waiting until the ducks leave their nests before taking the down is the hope that both parents and successfully fledged ducklings will return the next year. Paying a high price for cruelty-free eiderdown is worth it when it comes from a combination of happy wild ducks and human work carried out with patience.

A cloud of eiderdown placed on your hand will fill it with warmth, without feeling any weight. Materials science can explain this magic by structural properties at the microscopic level which make eiderdown feathers stick together and catch air. When female eider ducks line nests with their down feathers, they create a warm place, despite surrounding Arctic chill, for their eggs to hatch.

Where to see eider ducks
In the north of England, Scotland and Northern Ireland they can be seen all year round. Winter cold swells their ranks as many eider ducks overwinter in Britain rather than their Nordic breeding grounds, and some will travel as far south as Cornwall. Although ducks are largely associated with freshwater, eider ducks are rarely far from the sea. If you see a flotilla of beautifully chunky ducks on the sea in winter it is very likely to be eider ducks.
> St Andrews, Fife, Scotland
> Belfast Lough, County Antrim and County Down, Northern Ireland

Emperor dragonflies (*Anax imperator*)

> Seahouses, Northumberland, England
>   (where they are called 'cuddy ducks')
> Looe, Cornwall, England
> Biggar Bank, Walney Island, Cumbria,
>   England

Emperor dragonflies

Life divided by the surface of water

My lawn was in the clutches of ghosts.
Returning home to #NoMowMay after
a couple of months' pandemic lockdown
elsewhere, the grass was beyond knee-
high – perfect for dragonfly nymphs to
climb and await metamorphosis. I saw
one emperor dragonfly (*Anax imperator*)
emerge jewel-green and fly away, leaving
behind the translucent shell of its nymph
skin holding a stalk of grass.

Emperor dragonflies become even
more striking as some of their green
changes to equally brilliant turquoise
when they mature. With a body around
eight centimetres long and wings spanning
ten centimetres, they must be our most
magnificent insect, not only for their
bright colours but also their size. Yet in
the nymph form they inhabit for two to
three years underwater, they are not so
eye-catching – just drab brown, lurking
and hunting in still waters. Counting more
than a dozen nymph skins left on my lawn,
I realised that while I have blamed newts
for eating my tadpoles, the numerous
dragonfly nymphs living in the pond
have been consuming tadpoles in large
numbers. Although there were certainly
many emperor dragonfly nymphs growing
in my pond, adults don't share their water.
In order to make themselves attractive to
females, males are territorial over prime
egg-laying sites.

As adults, emperor dragonflies continue to be rapacious hunters, catching flying insects such as butterflies and smaller dragonflies, and eating them while continuing to fly, as they don't linger on perches for long. Despite the clattering racket of their wings if they blunder indoors or as they fly past, which can sound like something angry with a sting, people have no need to worry about them. Dragonflies only bite people defensively if they are caught. So if you look but don't touch, you will be fine. As well as flying high in pursuit of prey, emperor dragonflies can travel far. They have recently and rapidly expanded their range northwards from England, moving into Scotland.

Although emperor dragonflies continue to live near freshwater as adults, they never return to their underwater haunts. A couple of years ago, a friend was talking about their lost child, and being given comfort by the thought of their spirit being in a different element, kept apart for now by a barrier that was impassable. Writing while feeling the absence of my dog snoozing under my desk or watching me from his favourite places in the garden, I also think about missing loved ones. For many people, the story of the dragonfly moving from aquatic to airborne life is a comforting analogy.

Where to see emperor dragonflies
On the wing from May to October, and lurking underwater as nymphs year-round, emperor dragonflies inhabit still or slow-moving freshwater – ponds, lakes, canals and languid lowland rivers – where there is also plenty of vegetation in the water and at its edges. Tolerant of brackish conditions, they can also be found in seaside locations.

> Hythe End gravel pits, Berkshire, England
> Adventurers' Fen, Cambridgeshire, England
> River Stour, Foxearth, Essex, England
> Llangorse Lake, Powys, Wales
> Newly arrived in Scotland – look out for them in Galloway Forest Park

European eels
Awe-inspiring long-distance swimmers and kindred wild spirits
Seeing a fish out of water makes it hard to register that it is a fish. Walking the dog early enough for dew to be fresh on the grass around Virginia Water Lake, I saw a huge snake slide across the path. It was too big to be one of our native snakes and had a ridge along its back, and I realised that it was an eel. For eels on a mission, a short stretch of damp grass can be a pathway.

Although I often saw eels while

sitting on a riverbank looking into the water, it was years before I met an eel in its element. Duck-diving down and swimming along a riverbed, I came face to face with an eel winding its way through waterweeds.

In autumn the Sargasso Sea's siren call draws maturing European eels (*Anguilla anguilla*) from Europe back to their birthplace. As eels begin their journey downstream, their body turns from yellow to silver and their eyes become bigger.

They won't return to the rivers and lakes they lived in. Their purpose is to reach their spawning grounds of free-floating Sargassum seaweed to mate.

Swimming thousands of miles down rivers and across the Atlantic Ocean to the Sargasso Sea is one of the reasons that European eels are critically endangered. There is a high risk that they may become extinct in the wild. Dams, weirs and sluices obstructing rivers make it difficult for eels to travel from sea to freshwater, where they

spend years maturing, and hinders them when they seek to return to the Sargasso Sea to spawn. Another problem for European eel populations is that changing ocean temperatures may be changing the ocean currents that the freshly hatched eels ride from the Sargasso Sea to reach Europe.

The River Thames used to be so full of juvenile eels, called elvers, swimming upstream in spring that they were considered food for poor people, and fuelled the pie and mash shops of East London. Consumption is still contributing to the decline of European eels. Global demand for their delicious flesh gave 600,000 smuggled elvers intercepted at Heathrow airport in 2017 a market value of £1.2 million. Every eel that is eaten was once wild. All farmed eels are caught from rivers as elvers and raised in captivity, because no one has managed to breed European eels successfully. Since the 1980s there has been a ninety-five-per-cent decrease in the number of elvers joining adult populations in rivers and lakes.

Truly wild, with their defiance of captive breeding and their long-distance travels, eels can't be saved in zoos or aquariums – we have to stop eating them and keep their waterways accessible. Seeing eels winding through underwater plants in sinusoid waves, flashing silver as

< European eels (*Anguilla anguilla*)

their flanks catch the light, is a bewitching sight. I hope they will be there for the next generation to meet.

Where to see European eels
Swimming just above plants lining the bottom of a clear-running river or lake, or drifting on the surface is the best way to see an eel.
> River Stour, Kent, England
> Lough Erne, Northern Ireland
> River Thames tributaries, England
> River Dee, Kirkcudbrightshire, Scotland
> River Parrett, Dorset and Somerset, England

Gannets
Inhabitants of a liminal kingdom
Descending on the horizon and ending in watery detonation, northern gannets (*Morus bassanus*) mark their kingdom with dives that Red Bull cliff divers can only dream of. Dropping with speeds that can reach 100 kilometres per hour, they are driven 3 to 4.5 metres deep in the sea with a rapidity that makes it harder for the fish they hunt to escape. While performance divers feel the fear of height and impact, gannets make these descents hundreds of times every day.

Gannets have the form of birds but the engineering of cars, as their face, neck, stomach and back are protected by air sacs that absorb the impact of hitting water

at speed, and then immediately expel air, allowing them to sink down to the fish they seek. These responsive air sacs are just one of the features that makes them so adept at a hunting method that requires transitioning from aerial seeking of shoals to aquatic capture of individual fish. Gannets' eyes are positioned to allow them to calculate the angle at which their wings must fold a microsecond before hitting the water to create least resistance. In the air, the corneas on the surface of the eye focus on visual clues for the presence of fish far below them. Underwater, it is the lens of their eyes that provides the focusing, and their blue eyes become black as the pupils expand to let in as much light as possible.

In the high-speed mosh pit of a gannet feeding frenzy, there are collisions that are fatal, which is not the only reason why people should steer clear of gannets. Their serrated bills are not just good for hunting; life in a gannet colony is vicious. Their powerfully muscled necks are fortified for fighting as well as diving. Ferocity feeds successful colonies; a gannetry is a violent brawling pit where only the fierce survive. Accompanying the cost of living in a colony is the benefit of a hive mind that knows where shoals of fish are found. Our surrounding seas are divided between gannet colonies; each colony has its own fishing grounds.

< Gannets (*Morus bassanus*)

Where to see gannets

I first saw gannets after being startled by hearing the explosive bang of their plunge into the sea in the distance while I was swimming on a Pembrokeshire beach. It is well worth a look with binoculars to see the stunning combination of apricot-tinged heads with blue eyes. To be immersed in the action of their dives and brave the ammoniac stench of thousands of these birds together, take a boat trip to one of their colonies. More than half of the world's gannets live in the UK.

> Troup Head, Aberdeenshire, Scotland – a gannet colony on the mainland
> Bass Rock, in the Firth of Forth, Scotland – the biggest gannet colony in the world
> Little Skellig, off County Kerry, Ireland
> Great Saltee, off County Wexford, Ireland
> Grassholm Island, off Pembrokeshire, Wales

Great crested grebes

Satin bird

On a dull day, these are birds to cheer you up – they do a little dance, have fancy head plumes, carry their chicks around piggyback, and are a conservation success story. Great crested grebes (*Podiceps cristatus*) are a bird to lift your spirits.

Great crested grebes perform dives

for different reasons. Being fish-eaters, their primary reason for diving is to catch fish. Diving provides them with an escape route when startled. They also use a long shallow dive to get close to a potential mate and then make a grand appearance by suddenly rearing up out of the water in order to declare their intentions. If not rejected, the suitor can proceed with the object of its affections through a series of ritual movements. Great crested grebes use choreography and coordination of

head-shaking, turns and feather flicks as they move together, dancing across their lake's surface. Diving deep, they collect waterweeds and meet on the surface to display leafy offerings to each other.

After mating, great crested grebes take turns to sit on their eggs in their nest while the other partner goes fishing. Once the chicks are hatched, family life becomes mobile by making use of the broad back of adults. Although the chicks are able to swim soon after hatching, they don't have

the speed and stamina of their parents; hitching a ride on a parent's back allows the family to abandon the nest and move freely around their lake. But this only lasts for a few weeks. After a month, the parents split up and each takes a chick or two with them. Standard territorial rules then apply, and if the divided family meets they might even be hostile towards each other. February's romance does not last long for great crested grebes; perhaps this offers solidarity to anyone feeling jaded by Valentine's Day.

Mating glory nearly had a catastrophic impact on great crested grebes in Britain. At the Great Exhibition of 1851, Robert Clarke & Sons exhibited the feathered skins of four great crested grebes in full breeding plumage. Their soft feathers became in demand as a substitute for animal fur in boas and muffs. Britain's population of great crested grebes plummeted to just thirty-two breeding pairs. Women from the Plumage League and the Fur, Fin and Feather group campaigned against the use of great crested grebes in clothing, despite being mocked by men. One Royal Charter from Edward VII and many years later, these bold women left us a legacy which included the establishment of the RSPB and the continued presence of great crested grebes on our lakes. Now their alternative name

< Great crested grebes (*Podiceps cristatus*)

of 'satin bird', referring to the delicacy of their feathers, is no longer ominous.

Where to see great crested grebes
Great crested grebes find it difficult to get launched, so they inhabit lakes, slow rivers and gravel pits that provide them with a long runway rather than small ponds. In winter they are found in coastal areas.
> Stanborough Lakes, Hertfordshire, England
> Loch Leven, Perth and Kinross, Scotland
> Windermere, Cumbria, England
> Llyn Tegid (Bala Lake), Gwynedd, Wales
> Lough Neagh, Northern Ireland

Grey seals
Arabesque swimmers
Every time I've seen seals it's been too exciting to remember to look at their nose and decide if it is Roman. In distinguishing between the two species of seal that breed in UK waters, the common seal (*Phoca vitulina*) and the grey seal (*Halichoerus grypus*), noses are decisive. Grey seals have a long snout which appears Roman, in contrast to the short snout of common seals. It isn't often that you have the two species side by side for a nose comparison. Another way to tell them apart is that grey seals have almost parallel nostrils, while the nostrils of common seals meet at the bottom, forming a V shape.

Fully grown grey seals can weigh over 300 kilograms – an intimidatingly sized animal to meet in the water, especially when seals are inquisitive and use their mouths as well as flippers to investigate objects of interest, which can include swimmers. But being joined in the water by a seal is not something to fear. When the seals instigate interaction and they are not cornered, they will satisfy their curiosity and indulge in playful behaviours – more like a giant puppy than a scary beast. Avoid obstructing them, don't pursue them, and if they choose to interact with you, enjoy meeting one of our most charismatic mammals.

Taking a boat trip on the glass-bottomed boat at Gairloch on Scotland's west coast, I had a clear view of grey seals swimming in their arabesque style. Hefty seal bodies that seem ungainly when hauled out on a beach are transformed by the medium of water into powerful and elegant vessels. They spiral in flowing

lines, intertwine and extend through the water – a parallel to the grace and strength of outdoor swimmers I know who have a little bioprene – a nice layer of subcutaneous fat that works like neoprene. Their buoyancy and tolerance of cooler temperatures can give their winter sea swimming an appearance of effortlessness, in comparison to muscle-bound, pool-acclimatised athletes.

There is a time of year in which grey seals need space: their convergent breeding and pupping seasons. Pupping can be as early as September in Cornwall and as late as December in the Shetlands. Those limpid-eyed fluffy white pups look irresistibly cute. In their first three or four weeks of life, they are entirely dependent on their mother, who may abandon the pup if disturbed. No one wants to get caught in between a pair of randy 300-kilogram males competing over females. So it is best to stay away and avoid swimming at beaches where seals are hauled out during the breeding and pup season. Watch them from a distance, making sure you are not between them and the water, and use binoculars for a closer view. Our cool seas are home to about forty per cent of the global population of grey seals, a great example of not needing to travel abroad to see something a little bit special.

< Grey seals (*Halichoerus grypus*)

Where to see grey seals
> Monach Isles, Outer Hebrides, Scotland
> Falmouth, Cornwall, England
> Ramsey Island, Pembrokeshire, Wales
> Skerries, County Dublin, Ireland
> Farne Islands, Northumberland, England

Harbour porpoises
Shy swimmers

One of the few animals to possess an eponymous verb, harbour porpoises (*Phocoena phocoena*) may be seen porpoising around Britain and Ireland at any time of year. To porpoise is to move forwards while rising and falling. Subtle are the charms of harbour porpoises; they do not leap acrobatically. It takes a calm sea or a canny eye to spot them skimming along the water's surface – unless it is one of the rare occasions they are moving at speed, when like a hooligan on a jet ski they cast a spray of water in their wake.

Somewhat human-sized, harbour porpoises are up to 1.7 metres long and weigh about sixty kilograms, which makes them our smallest cetaceans. They inhabit areas of sea that are a wild swimmer's natural haunts – quiet, shallow waters. Capable of diving as deep as 200 metres, they mostly stay near the surface, emerging to breathe a couple of times a minute. They tend to stay away from boats. A gentle

Harbour porpoise (*Phocoena phocoena*)

swimmer blending in with the seascape has a better chance of seeing them than a speeding boat.

Although harbour porpoises are our most common cetacean, their populations are shrinking. Bottlenose dolphins have been observed and recorded on several occasions killing porpoises. It isn't clear why dolphins would practise porpicide, and suggested explanations include competition over food resources, generalised aggression and mistaken identity – male bottlenose dolphins kill calves and might have mistaken a porpoise for a dolphin calf. Humans have a bigger impact on porpoise populations than dolphins. Drowning when caught in fishing nets is one cause of porpoise decline. Human activities such as drilling, sonar and engines shattering the softer soundscape of the sea may be another cause of their dispersal, as porpoises are thought to abandon such noisy areas, in part because sound is so important to them as the way they locate food and detect predators. Porpoise echolocation is more sensitive than military sonar technology. Let us be grateful that under the EU Habitats Directive the UK and Ireland were required to designate protected areas for harbour porpoises. And also hope that, although decoupled from the EU, the UK continues to protect them.

It is a legal offence to disturb cetaceans, like harbour porpoises. This doesn't mean that you can't enjoy seeing them while swimming, but it is important to watch their behaviour for signs that your presence is disturbing them. If they keep moving away, changing direction when you change direction, diving if you pass near them and surfacing further away, these are indicators that they need more space between you and

Kingfisher (*Alcedo atthis*)

them. Perhaps then it's time to perch on the beach with a pair of binoculars and get a better view by watching them from dry land.

Where to see harbour porpoises

Harbour porpoises are present in numbers in our seas. Seeing them is more a case of becoming aware of their presence than searching for them. They eat small fish that gather in shoals – sand eels, whiting, sprat and herring. These shoals are also enticing prospects for seabirds. A flock of seabirds can be an indicator of a shoal of fish, and so a good place to look out for the little dorsal fin of a harbour porpoise just emerging from the water. They make use of tidal currents in their hunting for shoals of fish to dine on. Keep in mind that when harbour porpoise are active, currents may be running, so plan your swim – or not swimming but watching from the shore – accordingly.

> St David's Head, Pembrokeshire, Wales
> Barricane beach, Devon, England
> Portrush, County Antrim, Northern Ireland
> Loch Pooltiel, Skye, Scotland
> Dungeness, Kent, England

Kingfishers

Riding peaks and troughs

A splash and flash of colour in lowland waters that brightens the dullest day, kingfishers (*Alcedo atthis*) invest in numbers to overcome the odds against their species. Having up to six chicks per brood and three broods a season just about keeps our kingfisher population going.

Good times for a kingfisher are clear water with plenty of little fish and aquatic insects to eat. Bad times take many forms. Flooding in summer can turn river water turbid, which makes it difficult for them to hunt their food. If winter's cold forms impenetrable ice on the surface of water in their territory, they have to seek unfrozen water. If they venture into areas that are part of another kingfisher's domain, there is no sharing of space; fishing rights are fought over. In cold winters, kingfishers move to brackish and salty estuary and sea water that is slower to freeze. Occasionally they take advantage of garden ponds stocked with little fish or tadpoles. Moving to new places when winter is coldest and diving through holes in ice to catch fish might be why in German their name is *eisvogel*, meaning 'ice bird'.

Before you see a kingfisher, you are likely to hear its call, like a train or fast-moving car distorted by speed, as it flies low and swiftly over water. They return to favourite perches to hunt, and that is when a quiet watcher can see them at work. In breeding season, their frequent trips to small holes in riverbanks show you where their nest is, and this also tells you to avoid the area. Kingfishers are protected under Schedule 1 of the Wildlife and Countryside Act 1981 and it is an offence to disturb their nests. Breeding time is critical, as each chick needs to eat up to eighteen fish a day. Kingfishers are shy and can be deterred from feeding their chicks if a person is nearby. If you are swimming past and spot their nest, it's not a problem – just don't

linger. Kingfishers make new nests each year. With the nest at the end of a burrow they have dug into a riverbank, there is nowhere for the poo from all those chicks to go. By the end of the season, the nest is lined with little fish bones and scales.

If you catch sight of this brilliant turquoise and orange bird, give thanks to Emily Williamson, Etta Lemon, Eliza Phillips, Winifred the Duchess of Portland and Queen Alexandra. These five women founded what is now the Royal Society for the Protection of Birds to fight the Victorian fashion for plumage harvested by killing birds. It wasn't just exotic hummingbird or toucan feathers that were sought after – our native kingfishers' feathers and bodies were traded for ornamental hat trimmings. I am very grateful to see kingfishers decorating our riverbanks and ponds today.

Where to see kingfishers

> River Beane, Hartham Common, Hertford, Hertfordshire, England
> River Clyde, Motherwell, Strathclyde and Ayrshire, Scotland
> River Lagan, south of Minnowburn, Belfast, Northern Ireland
> Hampstead Heath Ponds, London, England (thanks to artificial bank built out of sand and cement for them to nest in)
> River Usk, south of Brecon, Powys, Wales

Lesser weever fish (*Echiichthys vipera*)

Lesser weever fish

Little fish that fell giants

Excruciatingly painful. Getting stung by a lesser weever fish (*Echiichthys vipera*) isn't a great start or finish for a swim, though it is an effective defence for a fish that is only about fifteen centimetres long. I remember walking into the sea thinking I could feel something sharp and not quite putting my foot down. Within a minute my foot hurt; after swimming, I hobbled home, had a day with a curiously numb foot and had a few weeks with a mark left on its sole. A friend who surfs said I must have stood on a weever fish but got off lightly by not standing on it properly. They had once experienced a full-on sting and crawled out of the sea with their eyeballs rolling.

Lesser weever fish are clumsy swimmers. Capable of short bursts of speed in order to catch crustaceans and small fish to eat, they lurk buried in sand with their eyes, mouth and venomous spine-tipped dorsal fin peeking out. Lying at the bottom of a pool seems impossible for the average person, though densely muscled swimmers or those with the breath control of free divers can manage it. Lesser weever fish manage the feat by lacking buoyancy – they do not have a swim bladder and do not float.

Encounters with lesser weever fish are more likely when walking in and out of the sea, at low tide and during warm weather. In summertime there are more people in the sea, and the fish also move into shallower water as the weather warms up. Although stings can be incredibly painful, most people do not suffer long-term effects, and there is an effective remedy. Because the venom is a protein, it is denatured and rendered inactive by exposure to heat. Immersing the stung body part in water that is at least 40 °C, but not hot enough to burn, will bring relief. On rare occasions, a person can have an allergic reaction or, if the tip of the spine breaks off and remains embedded in flesh, get an infection. In both cases, medical attention should be sought.

While the focus of a person who has been stung by a lesser weever fish is how to get relief from the pain, medical science has a different avenue of inquiry. Lesser weever fish venom is considered to have pharmacological potential. In laboratory conditions the venom has been found to cause the death of human carcinoma cells. This doesn't translate directly as a cure for cancer. It does add to understanding of biochemical triggers of cell change like programmed cell death, and there is a very low probability that it could become a biotech tool in the treatment of cancer.

Where to watch out for lesser weever fish
Lesser weever fish are native to the Eastern
Atlantic Ocean as far west as the Azores,
as far south as Morocco and as far north as
Norway, and in the Mediterranean Sea.

Swimming over a sandy seabed, you
might catch a glimpse of eyes looking out
of the sand, or a small dark triangular
shape that is the dorsal fin.

> Bournemouth, Dorset, England
> Freshwater West, Pembrokeshire, Wales
> Aldeburgh, Suffolk, England
> Rossbeigh, County Kerry, Ireland
> Newquay, Cornwall, England

Lion's mane jellyfish
Stinging beauty
Getting someone to pee on you is all very
well if that is your idea of pleasure, but
it won't take the sting out of a close
encounter with a lion's mane jellyfish
(*Cyanea capillata*). One creature stinging
100 people; being bigger than a blue
whale; the efficacy of urine for treating
stings: in the case of lion's mane jellyfish,
it is useful to sort myths from facts.

Their sting ranges from nasty to
serious. Lion's mane jellyfish are one of
the factors that make the North Channel
– swimming between Northern Ireland
and Scotland – so difficult. Guardians of

< Lion's mane jellyfish (*Cyanea capillata*)

the challenge, in the words of marathon
swimmer Kim Chambers, describe it as 'like
swimming through landmines'. If stung,
seek advice from health services because
lion's mane jellyfish venom can cause
systemic effects – not just localised damage
to the skin that was stung. However,
be aware that recommended treatment
protocols have not necessarily been
updated to incorporate recent scientific
research. In 2017 the Ryan Institute at
NUI Galway and the University of Hawai'i
at Mānoa published their finding that the
best first aid for lion's mane sting is to
rinse with vinegar, remove tentacles, and
immerse the stung area in 45 °C water or
apply a heat pack to it.

It is theoretically possible that one
jellyfish stung about 100 people at Wallis
Sands Beach in New Hampshire, USA in
2010. Up to 100 beachgoers were treated
for stings and the remnants of a dead lion's
mane jellyfish weighing nearly twenty
kilograms were recovered from the sea.
As it was breaking up, fragments of its
tentacles could have been distributed
through the water. Even when jellyfish are
dead, the stingers in their tentacles can
remain active for several days.

Individual lion's mane jellyfish can
be amongst the largest sea creatures in
the world. Alexander Agassiz, a naturalist
working as a professor at Harvard
University, reported finding one washed up

on a beach in 1865 whose dome was 2.3 metres wide, and whose length including tentacles was 36.6 metres long. This is longer than a blue whale. He used paces to measure the jellyfish, and even allowing for the vagaries of this measure it was sizeable. Most lion's mane jellyfish in British waters only get up to thirty to fifty centimetres wide, with proportionately shorter tentacles.

In the late 1800s, brothers Leopold and Rudolf Blaschka made glass models of animals and plants. Amongst their creations is a lion's mane jellyfish now owned by Cornell University. It is skilfully crafted, yet its static orderliness cannot capture the beauty of jellyfish pulsating in movement, tentacles a fine skein through the sea. In the open ocean, lion's mane jellyfish are an oasis of life. Fish that are resistant to their sting shelter amongst the tentacles for protection from predators. However, for some birds this is a strategic opportunity as they dive to catch fish concentrated around jellyfish, thereby minimising the time and energy spent hunting.

Where to watch out for lion's mane jellyfish

Their realm is cold water and the further north, the bigger they get, hence their alternate name, 'winter jellyfish'. If being worried about them impinges on your enjoyment of being in the sea, get swimming kit such as a rash vest and swim leggings that reduces the area of skin exposed to tentacles.

> Dublin, Ireland
> Culzean Bay, Ayrshire, Scotland
> North Channel, between Northern Ireland and Scotland
> Benllech beach, Anglesey, Wales
> Marsden, Tyne and Wear, England

Mayflies

Taking time to notice insects

A haze of mayflies dancing over water heralds the arrival of summer. We notice them more in their final life stage when they are airborne and fluttering their diaphanous wings, and miss their nymph life stage when they live underwater. Perhaps 'nymph' conjures up images of nubile water sprites – certainly not what I think about on seeing the bristly many-legged critters that are mayfly nymphs.

There are about 2,500 species of mayfly. The green drake mayfly (*Ephemera danica*) is common around unpolluted water in Britain. As nymphs they live in lakebeds and river bottoms where there is sand or gravel to dig into and form their burrow. They feed themselves by collecting particles of organic debris from water. They complete their life cycle in two years, or sometimes three years or one year. In all cases, most of their life is spent living underwater in their juvenile form of a

Green drake mayfly (*Ephemera danica*)

nymph. Their life on land is ephemeral. Entomologists named the genus Ephemera after the Ancient Greek word *ephemeros*, meaning 'lasting only a day'.

Mayflies' transition from water to air is achieved in two stages. When growing is complete, the nymph rises to the surface and hauls itself out of its skin, emerging as a winged subimago. It now has wings but no mouthparts, as it will never eat again. After a few hours resting on a leaf and drying, it sheds its skin again and emerges as an imago – the adult form. Male mayflies gather in a swarm and dance above the water. Dancing to attract a mate is the penultimate use of their energy. After copulating in flight, the male falls down exhausted and dies, and the female dies after dropping her fertilised eggs in water.

As nymphs, mayflies are preyed on by fish and carnivorous invertebrates. Resting subimagos are eaten by spiders and mammals. In flight, mature adults are consumed by birds. After mating, mayflies that fall on water are fish food again. Fly fishermen spend time and effort copying the appearance of mayflies to adorn fishing hooks in a manner that attracts fish accustomed to gorging on them.

Mayfly nymphs' transformation into dainty winged creatures isn't magic; it takes work eating, growing, shedding skins and moving from a watery realm to get there. Mayflies that hatch after a year of growth don't just have their lifespan halved; they are smaller and lay fewer eggs than those that take longer to grow. The shorter duration of the nymph stage is associated with increasing river temperatures. It is not yet known how changes in mayfly populations will impact the wildlife that feeds on them.

Where to see mayflies

Fly fishermen provide a location cue that mayflies should be on the wing in that river. Take advantage of long summer nights on weekdays with a riverside picnic watching mayflies dance in the light of the setting sun.

> Loch Awe, Argyll and Bute, Scotland
> River Teith, Callander, Stirling, Scotland
> River Teme, Ludlow, Shropshire, England
> River Bure, Norfolk, England

> Lough Sheelin, Counties Westmeath, Meath and Cavan, Ireland

Mute swans
Our biggest birds

Effectively arrested for his behaviour, the swan Mr Asbo is an extreme illustration of how people can be scared by a mute swan (*Cygnus olor*). For three years, rowers in the River Cam in Cambridge reported being attacked by a particularly aggressive swan.

He even capsized a lone rower and attacked him with his beak. Mr Asbo had supporters who argued that the swan had a right to be in the river – his home – that superseded the leisure pursuits of people, but conservators of the River Cam were granted a licence by Natural England to move him. Mr Asbo and his mate were captured and taken sixty miles away. His wings were clipped to prevent him returning.

Mute swans are particularly territorial when they are nesting or have young cygnets with them. They will swim fast towards a perceived threat, such as a rower or swimmer, hissing with raised wings. Weighing up to fourteen kilograms, mute swans can be intimidating. But Mr Asbo's behaviour is an extreme example. Usually if a person who has incurred the wrath of a swan changes direction and retreats, the interaction ends. It can mean using a different exit point from a swim or paddle than originally planned. Also, on seeing a nesting pair, an occupied swan nest or young cygnets, it is better to use a different section of river or lake to allow swans to be left feeling unthreatened during the couple of months when they are particularly sensitive to disturbance. It is also worth thinking about how busy a swimming or paddling spot is, and how your visit might be one of many.

< Mute swan (*Cygnus olor*)

Poor Mr Asbo the swan was trying to live on a stretch of river that has a lot of rowing, punting and swimming.

Ostensibly Mr Asbo was an example of human–wildlife conflict, except mute swans are semi-domesticated. They live in close proximity to people, occupying town ponds and rivers as readily as rivers and lakes in remote countryside. And for hundreds of years we have encouraged them to live near us by feeding them, for the convenience of having big and delicious birds on hand when we wanted to eat them.

Swans were a centrepiece for medieval banquets. In 1482 Edward IV passed the Act for Swans; yeomen and husbandmen were prohibited from keeping them. Swan ownership was not claimed by keeping them captive; instead they were permanently marked. Swans were free to roam, but notches and cuts on their beaks showed that they were private property. Swan marks were expensive to register, and this reserved swan ownership for wealthy people. Eventually, the New World introduction of turkey took hold, and domesticated turkeys became the festive meal. Now mute swans are protected under the Wildlife and Countryside Act 1981, which is why Mr Asbo could only be moved with a licence. Most of the time we can coexist happily with swans so long as we give them the distance from us that they are comfortable with.

Orcas

Apex predator of all oceans

The Tlingit, whose territory is in the Pacific Northwest coast of North America, have a legend about the origin of orca (*Orcinus orca*) that is in accordance with the absence of written records of anyone being killed by a wild orca. A hunter called Natsilane carved a black fish out of wood, which came to life when launched in the sea. He ordered the black fish to kill his fickle brothers-in-law, who had abandoned him on a deserted island. When this was done, he then instructed the black fish to never harm humans again. Blackfish, or orcas or killer whales, are revered in Tlingit culture as a powerful force of nature with an affinity for humans.

Orcas are consummate hunters with a wide and inventive repertoire of hunting

< Orca (*Orcinus orca*)

techniques. They spit fish onto the surface of water as bait for catching seagulls. Working in teams, they make waves wash over ice floes in order to push seals taking refuge on them into the sea. They catch great white sharks to eat their livers, which are a rich source of squalene, a precursor for synthesising hormones. Catching a person from a kayak or canoe would be well within the skills of an orca. Yet, while mythology of the Pacific Northwest attributes power and strength to orcas, they are not feared and it is considered auspicious to see them. A pod of orcas ramming boats off the coast of Spain and Portugal were thought to have been triggered into this defensive behaviour by one of their pod being seriously injured, causing the orcas to feel threatened.

Orcas have their own language of clicks, whistles and calls, and also imitate new sounds and use them in appropriate settings – i.e. they are one of the few species alongside humans to have vocal learning, a cognitive ability that underpins language. Living in matriarchal family groups, they travel together in pods consisting of extended family. Pods sharing similar calls are referred to as clans. They adhere to group traditions, from which fish they eat to behaviours such as greeting ceremonies. (Salish-Sea-resident orcas line up in two rows and then pile in to jostle each other – an orca moshpit.) It comes as no surprise, then, that captive orcas deprived of their rich cultural landscape are not representative of wild orcas' behaviour, and people have been killed by captive orcas.

Orcas are the aquatic mammal with the widest distribution. They are found in all oceans from the Aleutian Islands to Antarctica. Britain has a small resident group of orcas that live on Scotland's west coast. This group is unlikely to persist as they have not produced a calf for more than twenty-five years. Their reproductive failures may be due to polychlorinated biphenyls (PCBs) that were phased out of use from 1979 to 2000, as they are toxic to humans and animals but still contaminate European waters. In 2016 a killer whale from the west coast group died after being entangled in fishing lines. Levels of PCBs in her body were twenty times higher than the threshold at which physiological effects are caused. Most orcas seen in British waters are visitors from Iceland.

Where to see orcas
It might be more relaxing to see these near seven-metre-long, six-tonne apex predators from dry land. Or a large boat. May to July is the peak season for sightings in Orkney and Shetland.
> Duncansby Head, Caithness, Scotland
> Cantick Head, Hoy, Orkney, Scotland
> Noup Head, Westray, Orkney, Scotland
> Sumburgh Head, The Mainland, Shetland, Scotland
> Mousa Sound, The Mainland, Shetland, Scotland

Osprey
Fish hawk
Summer visitors that attract their own audience of tourists, osprey (*Pandion haliaetus*) are once again a barometer of our seasons. After overwintering in Africa, it is their habit to head north in late spring and settle in Europe to breed in summer. A standard triptych of peril for raptors – persecution, egg collection and demand for taxidermied adults – caused their numbers to dwindle until only one osprey pair bred in Britain in 1916, followed by none.

Why Britain got a second chance with osprey – a navigation error, storm wind or the sight of voluptuously enticing fish in a loch – no one knows for sure. But in the early 1950s osprey from Scandinavia tried to nest at Loch Garten. It only takes one egg collector to ruin birds' nesting attempts, and for several years eggs were stolen from Britain's sole osprey nest. In 1958 the RSPB set up twenty-four-hour surveillance of the nest at Loch Garten. A misty night provided opportunity for an egg collector to steal osprey eggs and replace them with chicken eggs coloured with shoe polish. In 1959 the gaggle of volunteers and RSPB staff lived in tents at 'camp osprey' and the osprey pair were

Osprey (*Pandion haliaetus*)

successful: three chicks hatched at Loch Garten. Since then, osprey have gradually expanded their British occupation. After 150 years of absence from England, osprey arrived and bred south of Hadrian's Wall in 1999.

Osprey are very amenable to humans doing nest-building for them; telephone poles can be just as appealing to them as trees for their sturdy constructions of stick exterior lined with bark, plants or turf. They also readily nest on artificial nesting platforms installed as part of conservation efforts.

Being exclusively fish-eaters, osprey are equipped to handle them. After diving down to catch fish in their talons, they fly off with the fish head pointing forwards – not so that the unlucky fish has a forward-facing view for the journey, but because it is more aerodynamic for the osprey's flight. Having a reversible outer toe makes adjusting fish position easier. Osprey also have built-in swimming goggles, technically a third eyelid, which are helpful when they dive underwater to catch fish.

Where to see osprey
If you are familiar with birds of prey, an osprey's 1.5-metre wingspan paired with a relatively slim body is distinctive in the way a supermodel stands out in a crowd of rugby players. These birds are sensitive to nest disturbance, but tree climbing isn't necessary to see them. Enjoy fleeting and distant views as osprey hunt across a lake. Set yourself up away from their nests so that you don't disturb them – in any case, being close to the nest of a large bird is a recipe for large bird poo landing on you. Get close-up views with binoculars when you are at a lake or loch, or be an armchair traveller and watch them on their nest via the RSPB's osprey webcam set up at Loch Garten.

> Bassenthwaite Lake, Cumbria, England

Otters (*Lutra lutra*)

> Rutland Water, Leicestershire, England
> Poole harbour, Dorset, England
> Glaslyn, Gwynedd, Wales
> Loch Morlich, Highland, Scotland

Otters

Loved and hated

We nearly annihilated otters in the British Isles with indirect weapons of destruction, including organochlorines as pesticides, habitat destruction and running them over accidentally, as well as the direct persecution of otter hunts. By the mid-1970s this species that was once widespread had dwindled to small fragmented populations. And then in 1978 otter hunting was banned. Use of organochlorine chemicals declined. Captive breeding and reintroductions increased the otter population. The progeny of otters that had survived in coastal areas when our rivers were blighted spread back inland. A couple of years ago on a swim trip in the Outer Hebrides, we gleefully took photos of a road sign that warned 'otter crossing'. Otters are now back in every county, and even appearing in Birmingham's canals.

Don't read *Tarka the Otter* or *Ring of Bright Water* – they don't have happy endings, though the films made of these books helped to rehabilitate the image of otters. They earnt their vermin status by being apex predators that dine on wild fish and fish-farm stocks. For most people, it is only in films that otters can be seen in their fluid swimming glory. Primarily, they swim to hunt. When food is abundant, otters have time and energy to play. For example, they slide down steep banks into water, which is practical but some otters do it more than necessary – it is not only locomotion but also play.

In south-west Turkey there is a crystal-clear river I love to swim in. It is plump with shoals of fish – ideal for otters. Someone told me that my friend Umut knew when to see the village otter. I did not know the Turkish for otter, but as we both knew their scientific name– *Lutra lutra* – he understood what I wanted to know. He told me that she came to the bridge at 10 p.m. My companion and I sat with a bottle of wine and watched the water. She was on time. First an advance ripple in the water and then we could see a fluid dark shape and hear soft splashes. She was typical for an otter. She had her territory and her routine, and was a delight to watch.

Where to see otters

Otters are memorable, so people following a range of pursuits from dog walking to canoeing may know local otter places. Myles Farnbank, Head of Guides and Training at Wilderness Scotland, told me, 'River-based otters are mainly nocturnal. They have two periods of activity: the first is four to six hours after dusk, the second just before dawn when the otter retires.' He also said that coastal otters are a bit easier to spot when the days are shorter as their hunting is concentrated in daylight hours. You can look out for spraints – essentially a type of poo that

otters use to mark their territory. Or when you see those 'otter crossing' signs on roadsides, make a plan to come back to the water quietly and wait for otters.

> River Stour, Dorset, England
> River Bure, Norfolk, England
> Glenarm River, County Antrim, Northern Ireland
> North Uist, Outer Hebrides, Scotland
> Loch Torridon, north-west Highlands, Scotland

Pike

Aquatic wolves

The poet Ted Hughes cast pike (*Esox lucius*) as tigers when he described their markings. Young pike are striped with gold that breaks up into speckles as the fish matures. Like us, Ted Hughes inhabited an era when tigers are part of our repertoire of fearsome creatures. But familiarity with pike here is older than knowledge of creatures acquired from distant places. An eleventh-century Bavarian fairy-tale calls pike 'wolves among fishes'. In twelfth-century England, pike were called aquatic wolves, and a book on fishing published in 1577 called them freshwater wolves. Britain's last wild wolf was reportedly shot in 1680, although rumours of wild wolf sightings persisted until 1880. With wolves gone, are pike our wildest remaining predator? And of course a question for swimmers is: do we need to worry about being bitten?

Fishermen are sometimes bitten by pike when they have reeled them in and are detaching them from the hook or holding the pike – not an unreasonable position from which to administer a bite. Swimmers don't need to worry. Considering that most lakes and slow-moving rivers have pike in them and the tabloids have not been plastered with tales of killer pike, it seems that swimmers can cohabit with pike without losing fingers or toes.

Pike will take prey that are up to about half their size. The longest pike recorded was 152 centimetres and weighed twenty-eight kilograms; the heaviest weighed thirty-one kilograms but was only 147 centimetres long. This makes a swimmer more than a mouthful, and pike stick to eating mostly fish and the occasional waterfowl. Pike hunt by stealth, cruising slowly and nonchalantly until they are within striking distance, which is usually within a metre of their

prey. Using power from their tail to straighten their streamlined body from a compressed position forwards through the water gives them an acceleration of up to a g-force of 15. A Porsche only has acceleration of about 1.24g. Robotics engineers have taken inspiration from pike and their phenomenal accelerative power to design robots that can navigate turbulent water. So swimmers looking for a fast start in a race or burst of speed in the water might do well to think pike.

I've watched pike from the riverbank on sunny days when the water is clear: their dappled flanks catch the light, and their sharp jaw distinguishes them from any other river fish. Once I met a pike while swimming. It was a leisurely encounter. My niece spotted it first and pointed it out. We hung in the water watching each other. Then slowly it swam past us and disappeared amongst waterweeds. We had been thrilled to see it; the pike seemed pretty disinterested.

Where to see pike
In March and April when pike are spawning they congregate at the shallow margins of rivers and lakes where there are waterweeds to lay their eggs.
> Stoney Cove, Leicestershire, England – like shooting fish in a barrel. At this diving centre, which also admits swimmers, pike are habituated to the presence of people and the spring-fed water provides good visibility.
> River Wye, downstream from Hay-on-Wye, Powys, Wales
> River Stour, Suffolk and Essex border, England
> Loch Lomond, Loch Lomond and the Trossachs National Park, Scotland
> Lough Derg, County Donegal, Ireland

Porbeagles
Not a threat but threatened
Being mistaken for a relative is unfortunate when that relative is the great white shark. Porbeagles (*Lamna nasus*) don't have the sensational and often fictional media profile of their larger relative the great white. But since porbeagles are indeed sharp-toothed sharks, perhaps the first thing a person wants to know is, who gets bitten by them? The answer is fishermen.

With ninety per cent of their diet consisting of bony fish and their second most consumed food being squid, porbeagles are not looking to eat people. Occasionally, some sharks bite people when they mistake surfers for seals. But not porbeagles, since they don't eat seals or other marine mammals.

If you consider being hooked on a fishing line and landed aboard a boat or

< Pike (*Esox lucius*)

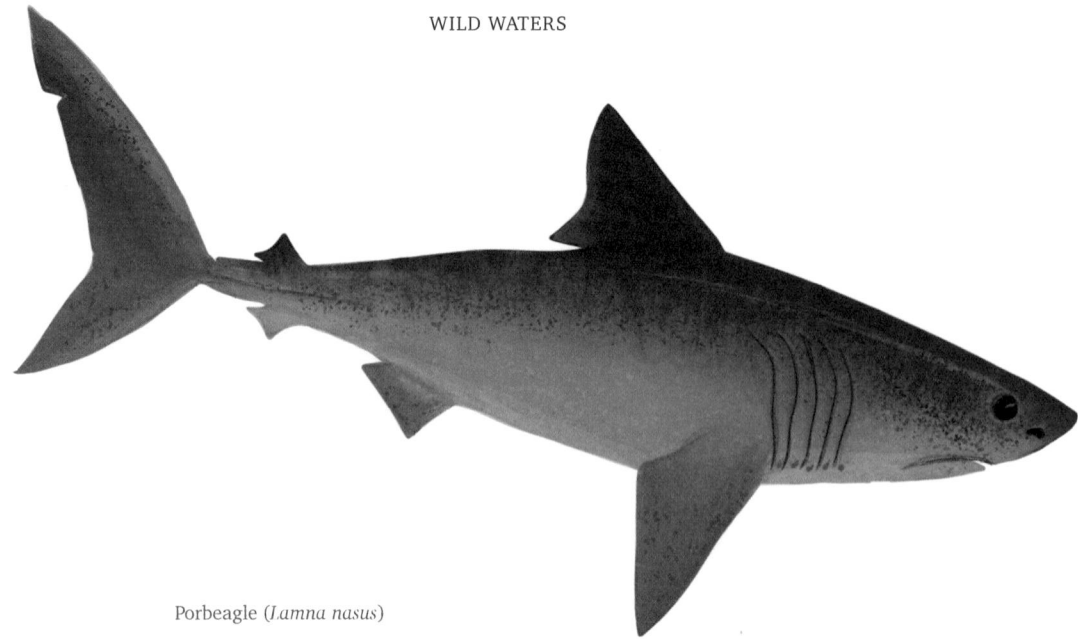

Porbeagle (*Lamna nasus*)

dragged to the surface at the side of a boat to constitute provocation, porbeagles aren't known for unprovoked attacks on humans. Porbeagle bites on humans tend to happen when fishermen are trying to release them from fishing hooks or when landed porbeagles are thrashing on a boat deck. Porbeagles are caught as bycatch, for example in longline fishing for tuna; they are also targeted by recreational fishers. Although the recreational catch and release of porbeagles is legal, the sparse evidence available on shark survival after being released indicates that a significant proportion of them die. It is a confounding arena in which the process of getting data might cause additional stress to sharks. Additionally, angling organisations run shark-tagging programmes in which catch and release is cast as a shark conservation tool. With their population declining due to commercial and recreational fishing,

porbeagles are likely to become an endangered species.

Where to see porbeagles

In summer, porbeagles come inshore and towards the surface. This vertical movement is led by their prey: bottom feeders in winter, shoals of mackerel in summer. Although porbeagles are found in the northern Atlantic and the southern oceans, they are not present in the tropics. They inhabit a narrow range of water temperature from 9 °C to 17 °C. Using an internal heat exchanger, capturing warmth from metabolic processes in their muscles, porbeagles keep their brains and eyes warmer than the seawater around them. Outdoor swimmers accustomed to using a silicon swim hat to reduce brain freeze will understand the advantage of warm brains.

Offshore oil platforms can sometimes host shivers of porbeagles. In 2014 at

Queen scallops (*Aequipecten opercularis*)

least twenty porbeagles spent a few days swimming around an oil platform, cruising at surface level with their dorsal fins emerging from the sea. Is it reassuring to point out that there are certainly porbeagles in our coastal waters and swimmers and surfers haven't had problems with them? Porbeagle hotspots tend to be offshore in deeper water, and they are enticed towards people by bait on fishing rods, not the prospect of dining on people. With a name arising from the Cornish for port (*porth*) and shepherd (*bugel*), porbeagles remain a living element of Cornish language.

> St Ives, Cornwall, England
> Dunnet Bay, Caithness, Scotland
> Milford Haven, Pembrokeshire, Wales
> Galway Bay, County Galway, Ireland
> Whitby, Yorkshire, England

Queen scallops
Swimming and seeing molluscs

For now England, Wales, Scotland and Northern Ireland form a United Queendom, so perhaps we should celebrate queenies. Europe's largest fishery for queen scallops (*Aequipecten opercularis*) is in the UK. Queen scallops are smaller than king scallops, but queenies are better swimmers. Mostly resting on the seabed while filter-feeding, scallops swim to migrate and also to escape when they see you coming. Yes, scallops are shellfish with eyes. Their ability to detect shadow and movement from visual cues is not just useful for predator evasion. Scientists have proved that the structures dotting their edge that look like eyes really are eyes by playing scallops movies of drifting particles of food. As the food particles loom closer in the movie, scallops open up their shells ready to feed.

Queen-scallop-style swimming is not an effective way to escape people. They jolt up from the seabed in jerky movements, using their muscular body to open and close their shell. Then they drift back down to the seabed only a short distance from where they started. It is an effective way to escape starfish, which are their main predators. Being unable to escape swimmers and divers is overall a net benefit. It means that in our lust for their strangely sweet flesh we can consume sustainably caught scallops. While being hand-caught is unfortunate for an individual scallop, overall it offers a sustainable fishing method, unlike dredge fishing for scallops, which destroys seabed habitats.

Bountiful scallop spawning offers a means of farming scallops. Abundant release of spermatozoa by male scallops and ova by female scallops results in an abundance of fertilised ova, leading to vast numbers of tiny immature scallops – far more than there is space for in established colonies. Nets left hanging near colonies around spawning time become covered in slimy grit, which is actually immature scallops and their accompanying sticky threads of byssus. Immature scallops caught on these nets can be moved into aquaculture where they are farmed in nets to reach eating size, while leaving plenty of wild scallops in the sea.

Where to see queen scallops
Along the shoreline, you can often find scallop shells washed up. Their orangey– red tones marbled with white are a pop of colour on winter beaches. To see queenies intact, you need to duck down to the seabed. While they are lying still, they can be camouflaged by small seaweeds and sea sponges that grow on them. It is more likely that your attention will be drawn by their movement in a burst of jet-propelled swimming. As filter-feeders, they benefit from currents moving water past them, so if you are thinking of checking them out, pay attention to tides and aim for the slack tide.

> Isle of Man – especially in the years they run the Queenie festival.
> Lamlash Bay, Isle of Arran, Scotland – a no-take zone created by the local community and subsequently given legal protection, so don't be tempted to forage here.
> South Bay, Scarborough, North Yorkshire, England
> Lyme Bay, Devon, England
> Elly beach, Blacksod Bay, County Mayo, Ireland

Reedmace (*Typha latifolia*) >

156

Reedmace

Year-round abundance

At any time of year, reedmace (*Typha latifolia*) offers food, in contrast to most plants, whose bounty is limited by seasons. Autumn is the perfect time to get to know this edible plant. In this season, leaves are topped by distinctive seed heads. There is nothing else that looks like five-foot-tall grass topped with fat sausages on sticks except its closely related and equally edible relative, lesser reedmace (*Typha angustifolia*). In autumn, as plants hunker down for winter dormancy, many of them pack their underground parts with the products of summer sunlight. Reedmace fills its rhizomes with starch and they are delicious roasted. But be prepared for messy work to get them.

Reedmace grows in mud at the edge of ponds, lakes and slow-moving rivers. It loves the mucky bit at freshwater edges. Before you can collect its rhizomes, you should have permission from the

landowner to pick them. A quirk of British law is that while you legally can pick above-ground parts of most plants without landowners' permission, picking below-ground parts requires permission from the individual or organisation that owns the land. Reedmace is vigorous and common, so sometimes as part of land management it is cleared from pond and river edges, a case in which a hungry wildlife enthusiast can help. More often than not, landowners, when asked nicely in advance, are OK with people exploring how nature tastes. Reedmace rhizome-harvesting season runs from autumn until spring, when it moves its stored energy into sending up shoots of growth.

At first these shoots are small, pointed tips. Growing quickly, they become cylinders like leeks or heart of palm. When well washed, they can be eaten raw. Or chop them finely as an addition to soups and stews, or chuck them whole on a barbecue. Reedmace shoots stay crunchy and tasty until the middle of summer, when rigid flower spikes start to form in their cores.

Perhaps reedmace is least exciting as a food in midsummer, when all it offers is an abundance of pollen. It is fairly easily collected by tapping the flower heads while holding them in a bag to capture the big yellow clouds they readily yield. Not just looking like yellow flour, reedmace pollen can be used in baking as a substitute for flour.

Caution is needed to be sure to avoid collecting rhizomes or shoots from poisonous yellow flag iris instead of reedmace. Those distinctive reedmace seed heads are a great introduction, but you should also learn how to recognise yellow flag iris so that you don't mistakenly eat it. Another caution arises from reedmace's ability to clean aquatic landscapes. It is able to absorb pesticide residue and heavy metals from contaminated water and soils, so you only want to harvest reedmace for eating in places where the water and underlying soil is unsullied.

Where to see reedmace
Pretty much any body of slow-moving or still water in lowland areas.
> Lough Neagh, Northern Ireland
> Frensham Great Pond, Surrey, England
> Aberlady Bay, East Lothian, Scotland
> Gormire Lake, North Yorkshire, England
> River Wensum, Norfolk, England

Salmon
Autumn's acrobats
Being rammed by a few kilos of salmon is a bit of a surprise while on a refreshing morning swim after a party night. If Keith had had reflexes quick enough to grab it – and a fishing licence – we would have had salmon for supper. Salmon can be territorial,

Atlantic salmon (*Salmo salar*)

and are also predators that strike fast at a potential meal. Later we met a gillie – a hunting and fishing guide – who told us that the yellow camera float dangling at the back of Keith's speedos was of a size and colour to lure salmon. It is most likely that the salmon took the float for a target and didn't notice the rest of Keith in the tea-coloured water of the River Feshie.

Atlantic salmon (*Salmo salar*) grow large by feeding in the richness of the sea, but use rivers as a lower-risk site for laying their eggs. Travelling back to their riparian spawning grounds is the cause of one of our wildlife spectacles – leaping salmon. They can launch themselves up to three metres out of the water if there is a deep

enough pool in front of the obstacle they are seeking to pass. It takes the energy of sprint muscles running the length of their bodies to hurl themselves up in the air.

They bury their eggs in gravel, and juvenile fish emerge after 40 to 145 days of incubation, depending on water temperature. Swimmers enjoying salmon rivers should take care to avoid disturbing areas of gravel from late autumn to late spring, as the stability of the gravel in which eggs are laid is essential for embryos to develop and emerge as juvenile fish. When they have grown to about twenty centimetres long, they migrate from rivers to the sea, mostly by cover of darkness. Salmon may return to the river they hatched in after one winter at sea, weighing a couple of kilos. But salmon that spend more than one winter at sea grow even larger and heavier, with females measuring up to 1.2 metres long and weighing twenty kilograms, while males can reach lengths of 1.5 metres and weigh thirty-six kilograms. Salmon return to the river they hatched in to spawn. They use the Earth's magnetic fields to navigate across oceans and seas to the coastline they originated from. They then use olfactory cues in freshwater to find their natal spawning ground.

Where to see salmon
In Scotland, Sunday is a good day to go in salmon rivers and avoid the ire of people who pay a lot for salmon-fishing licences, as salmon fishing is prohibited on Sundays. Seeing leaping salmon is most likely from mid-October to mid-November. Salmon that have reached their spawning ground early lurk near riverbanks, an ideal spot to look for them in their magnificent spawning hues of crimson for the males and purple for the females.
> Falls of Shin, River Shin, Sutherland, Scotland
> Stainforth Force, River Ribble, Yorkshire, England
> River Tamar, Devon, England
> Cenarth Falls, River Teifi, on the boundary of Ceredigion and Carmarthenshire, Wales

> River Finn, Donegal, Ireland and County Tyrone, Northern Ireland

Sea hares
Sluggish but sophisticated
In springtime, sea hares (*Aplysia punctata*) form an orderly queue for their orgy. They are hermaphrodite – they have both male and female sex organs – and deploy that characteristic in mass mating events that take the form of stacks of sea hares. At the bottom of the pile the sea hare is fertilised, and at the top of the pile the sea hare is the fertiliser. But throughout the rest of the stack, every sea hare is both fertilised and fertiliser. If you miss their mating spree, you may see where there has been one from the strings of pink–purple spawn left on seaweed.

Despite their name, sea hares are molluscs and look more like slugs than hares. Concealed beneath rippled flaps on their backs is a shell. While the rhinophores on top of their head were thought to resemble hares' ears, they function like a nose as a chemosensory organ, and don't detect sound. Another set of tentacles at the front of their head may also be used in chemoreception, like the rhinophores, and possibly for mechanoreception – responding to

Sea hares (*Aplysia punctata*) >

mechanical pressure and thereby assisting in movement.

Sea hares move more by crawling and floating than by swimming between the seaweeds they eat and between domains. Juvenile sea hares tend to live in deeper water. They return to shallower water as adults when they breed. Being slow-moving targets without a turn of speed or penchant for hiding, and growing up to twenty centimetres long, they seem like an easy mouthful for predators. However, sea hares have two chemical weapons at their disposal that act as deterrents to anything trying to eat them. Sea hares can release a red–purple dye when disturbed. Rendering the water coloured and opaque makes it harder for predators to get hold of them. Sea hares synthesise this dye from red pigments in the seaweed they consume. Another chemical at their disposal is opaline. They squirt this sticky substance at predators. It contains an assortment of chemicals, which seem to deter predators

by two means. By blocking the sense of smell, it reduces appetite. It also sticks to predators and takes a long time for them to clean off, allowing the sea hares time to move away.

Where to see sea hares
Since sea hares don't need to hide or swim fast to avoid predators, they are really easy animals to watch as they go about their business without being disturbed by the presence of observers. Springtime is when they congregate in shallow water to breed, which concentrates them in easily accessible places to watch them. Rock pools big enough to swim in are perfect. Rocks sheltering water from waves are conducive to clearer views. Being in the water allows for the interesting experience of being looked in the eye by a wild animal that doesn't feel the need to flee.

> Runswick Bay, Yorkshire, England
> Constantine Bay, Cornwall, England
> Strangford Lough, County Down, Northern Ireland
> Porth y Pwll, Anglesey, Wales
> Kingsbarns beach, Fife, Scotland

Seven-armed starfish
Fragile strength
Social conditioning tells us that fragility is a negative trait and loss should be mourned. Seven-armed starfish (*Luidia*

ciliaris) teach us otherwise, as one of their strengths lies in fragility. Their limbs are brittle and snap off, allowing them to regenerate new limbs to replace ones damaged by predators. Last time I saw a seven-armed starfish, it was temporarily a six-armed starfish with a gap waiting for an arm to regrow. Even when they have all seven arms, you can often detect the pattern of previous loss, as the newer ones are shorter. Despite the stories that missing limbs tell, nineteenth-century naturalists were desperate to get seven-armed starfish, as their fragility and propensity to drop all their arms in response to being caught made intact specimens rare.

Seven-armed starfish have unusually long feet for a starfish. Hundreds of them line the undersurface of each arm. With the tips of their arms curled upwards so that they don't drag, these extra-long feet enable them to walk rapidly across the seabed. When the starfish rise up into a walking position, some of their prey start taking evasive action. Occasionally seven-armed starfish have been seen out of the sea hurrying across damp sandy beaches. Exposed on a beach, they are vulnerable to attacks by birds. They can also be seen underwater in hot pursuit of their prey – brittle stars, urchins and other starfish. Instead of suckers, the seven-armed starfish's feet end in tiny knobs. Starfish with knob-endings can walk faster than those with

Seven-armed starfish (*Luidia ciliaris*)

disc-ending feet (or suction pads).

Arms are vitally important to these starfish. It is where they keep their feet and gonads. In the middle is their stomach and mouth. Sometimes seven-armed starfish entirely engulf their prey in their mouth. They can move and if necessary rupture the plates of their mouth in order to swallow their food whole. Although seven-armed starfish aren't the most colourful of our native species, they occupy a particular niche in the web of life. When they are abundant, brittle stars disappear. This is thought to be helpful in sustaining populations of sponges that are consumed by brittle stars.

With arms and feet but no eyes or bones, starfish are not in any way a fish. Pedants are starting to call them sea stars.

Where to see seven-armed starfish
Find them in rock pools, little and plump with chubby arms, and barely as big as a 10p coin. Out in deeper waters, they mature and grow much bigger. Their limbs morph into long elegance and they can be up to sixty centimetres across. But these large individuals loiter 50 to 100 metres deep where there is an abundance of prey, below the depths at which most seaweeds grow and deeper than swimmers reach, though they are very familiar to British divers.

> Lamlash Bay, Isle of Arran, Scotland
> Runswick Bay, Yorkshire, England
> Abersoch beach, Gwynedd, Wales
> Carnivan beach, County Wexford, Ireland
> Lannacombe beach, Devon, England

Short-snouted seahorse (*Hippocampus hippocampus*)

Short-snouted seahorses

Charismatic chimeras

Seahorses are little critters that drive conservation schemes that benefit more wildlife than just themselves; what is more intriguing than a fish with a monkey's tail, a horse's upper body, and a kangaroo pouch? A successful campaign for a review of the government's decision to allow a temporary oil rig in Poole Bay in spring 2019 was propelled by the Seahorse Trust's concern about wildlife affected by the rig, including short-snouted seahorses (*Hippocampus hippocampus*). To be honest, I am in support of oil extraction in the Poole Bay area, in no small part because we have a legal system where perceived violations of safeguards for environment and wildlife can be challenged. Conservationists in other countries are too often silenced by violence. At least 331

environmental defenders, most of whom worked in defence of land, environmental rights and Indigenous peoples, were murdered in 2020. Of these killings, 177 – more than half – occurred in Colombia. Farmers and activists there have blocked development of the oil market in Colombia. They are concerned about the damage fracking may cause to one of the most biodiverse countries in the world.

Christiane O'Mahony used seahorses as a way to examine challenges of modern womanhood in Ireland in her play *Seahorse*, prompted by seeing seahorse lifestyle as radically feminist, because female seahorses implant their eggs in male seahorses and give males more of the burden of child-rearing. Not that female seahorses abandon their breeding partners. Although a significant part of a seahorse's day is spent being stationary, and additional time is spent feeding and swimming, about six per cent of their time is spent specifically on greeting behaviour or courtship. They have several ways in which to make these social displays, most of which occur just after sunrise. Wrapping tails around each other, swimming together, quivering next to each other and twirling around an object are mutual activities. They also brighten as they approach each other.

Seahorses are protected by Schedule 5 of the Wildlife and Countryside Act 1981. A licence is needed to carry out activities that might disturb them. Surveying, filming or photographing them is considered to have potential for causing disturbance to them. It is illegal to touch them without a licence, and taking photos of them using a flash is banned by the Marine Management Organisation. It is also illegal to chase one, so don't swim after a seahorse if it swims away from you.

If you are lucky enough to spot one, keep in mind that hovering above them is also likely to cause them stress, as they can think you are a predator. They will turn their back on you as a defensive reaction – don't swim round them and try to look at them head-on. If a seahorse's colour darkens and it bends its head down, you are causing it stress and you should take this as a prompt to swim away.

Seeing seahorses

Seahorses are hard to see as they are rare and good at camouflaging themselves, and seeking them out is considered to be an activity that may disturb them. So, while I would usually highlight places where wildlife can be seen, in this case let's leave it as delightful happenstance if you see seahorses. On the other hand, reporting sightings of seahorses to the Seahorse Trust helps them to build knowledge of where they live and might need their habitat to be protected. So if you are out at sea and incidentally see one, or you find a dead one washed ashore, let the Seahorse Trust know where you saw it.

Smooth newts
Two kinds of life

Back in the days of killer newts the size of cars, swimming for fun wouldn't have been a priority. But *Metoposaurus algarvensis*, the fearsomely toothy and giant ancestor of newts, rampaged in the Triassic period, long before humans or anything resembling our ancestors existed. When you see tiny-toothed smooth newts (*Lissotriton vulgaris*) small enough to sit in the palm of your hand, you are looking at creatures with ancient lineage, older than dinosaurs.

Amphibian means 'both kinds of life'. Smooth newts have two separate stages of life in two different environments:

Smooth newt (*Lissotriton vulgaris*)

as tadpole-like larvae restricted to life in water, and then as newts that can go on land. Adult smooth newts oscillate between terrestrial and aquatic life. However, occasionally smooth newt populations exhibit two forms of maturity. Some individuals are paedomorphic: they never progress to terrestrial dwelling and remain in the form of aquatic larvae but reach sexual maturity; while others follow the standard progression from immature water-bound larvae to mature newts that can go on land. Having two forms of maturity provides an adaptive strategy that allows smooth newts to take advantage of auspicious living conditions. When water bodies are rich in nutrients and free of predators, it is advantageous to remain in aquatic larval form. But if

the aquatic environment becomes more hostile – for example, if fish are introduced – paedomorphosis disappears and mature newts are all terrestrial.

Amphibian skin is thin, unprotected by scales or fur or feathers, and also an organ through which they can absorb both oxygen and water. Smooth newts, even in their terrestrial form, do not need to drink as they absorb moisture through their skin. In breeding season, around the end of February and early March, male smooth newts develop magnificently patterned skin on their underside – dark spots speckled over yellow to orange. They also grow crests for the purpose of attracting female newts. Growing a crest takes energy, making it a proxy for fitness. Another technique is using their tail to move water loaded with enticing secretions towards females. While male newts expend energy on attracting females, the females make the effort to wrap each single fertilised egg in a leaf of an aquatic plant.

Where to see smooth newts

Smooth newts are a reward for moonless nights. When the moon is not bright, they are more likely to leave breeding ponds to seek food on land and move to different ponds. Moving when it is dark makes it less likely for them to be caught by a predator. Use a torch to spot them more easily amongst plants at the edge of water; whereas in daylight hours, your shadow passing over them will prompt them to hide.

> Natural swimming pools – plenty of plants cleaning the water and no fish or chemicals make these perfect abodes for newts.
> Hampstead Heath Ponds, London, England
> Cwm Llwch, Powys, Wales
> Loch Lomond, Loch Lomond and the Trossachs National Park, Scotland – swim in the loch and look for smooth newts in the ponds scattered throughout the surrounding national park.
> Upper Lough Erne Lake, County Fermanagh, Northern Ireland and County Cavan, Ireland

Snakelocks anemones

Tasty tentacles

Like people, snakelocks anemones (*Anemonia viridis*) contain multitudes, in the form of microbial communities living within them. Inhabiting the body of an animal that can defend itself with stings provides zooxanthellae, tiny algae, with relatively safe accommodation. In return, the host benefits from a share of the energy they capture from sunlight in the process of photosynthesis. It is a mutually beneficial arrangement in which food is exchanged for housing.

Snakelocks anemones are sedentary and live attached to rocks, kelp blades or seagrass, making their stinging defence critical for deterring animals that would otherwise consume them. As the tentacles defend themselves, they are rarely retracted, and are left lissomly swaying in water currents. Easily brushed against, they are sometimes noticed by swimmers after swimming as a red itchy rash on exposed areas of skin. It is worth taking a close look at the tentacles of snakelocks while in water, as this is the only place to find one of our most beautiful creatures. Anemone prawns (*Periclimines sagittifer*) are found exclusively inside snakelocks anemones. These tiny prawns have translucent bodies marked with indigo, white and lilac dots and stripes.

Snakelocks anemones take two different forms. Some are creamy–brown throughout their tentacles. Others have green tentacles that are pink-tipped. This green colouration is due to the presence of green fluorescent protein, which absorbs high-energy UV light from the sun and emits it as low-energy green light.

In the years after the Spanish Civil War, when food was restricted by government regulation of agricultural production and food rationing, snakelocks anemones were used as a food of necessity. Now *ortiguillas de mar* – literally 'little sea nettles' – is a dish that appears on many restaurant menus in the coastal area around Cádiz. Although Sardinia is an island with limited nautical tradition, snakelocks anemones are also a traditional food there in the form of *orziada*. As an animal that mostly dwells in water less than twelve metres deep, it is a seafood that can be accessed without the need for a boat or sailing skills. In both Sardinia and Spain, snakelocks anemones are dusted with flour and deep-fried before eating them.

If you are tempted to taste snakelocks anemones, remember that cooking is needed to render their sting harmless, and

Snakelocks anemone (*Anemonia viridis*)

they must be collected from water that is clean, not contaminated with sewage overflow or other factors that reduce water quality. There are also constraints exerted by responsible foraging. They should not be taken from marine reserves, and don't pull them off the rock or seaweed they are attached to unless you are certain of their identity. Only collect them if they are so abundant that your taking them does not decimate the population you see. They don't keep, so you need to have cooking facilities nearby. These limitations might encourage you to stick with just looking at them.

Where to see snakelocks anemones
> Osmington Mills, Dorset, England
> Porth Colman Cove, Gwynedd, Wales
> Loch Dhrombaig, Sutherland, Scotland
> The Silver Strand, Donegal, Ireland
> Bouley Bay, Jersey, Channel Islands

Sweet flag
Inconspicuous exotic
Green-flowered, green-leaved, blending into water edges, and visually outshone by more flamboyant-looking plants, sweet flag (*Acorus calamus*) is easily missed. But if you get to know this humble plant, you may observe it with awe and delight as you understand its ecological sway and medicinal power, and appreciate its flavour.

Sweet flag is a waterside plant with global reach. My sweet-flag revelation happened twice. Talking with Dr Michelle Baumflek about her work with Indigenous communities in Maine, USA was the first time I paid attention to this plant. It is not only an important component of traditional medicine for Indigenous peoples in North America. It features in Ayurveda and traditional Chinese medicine. Some of its uses are being validated by contemporary scientific trials. In northern Maine, sweet flag, locally called muskrat root, may be one of the most widely used medicinal plants. Maliseet gatherers rely on specific sites to collect it. When a car park was built on the primary Maliseet muskrat root collection site, tribal resource managers wanted to identify new collection sites. Working with the Houlton Band of Maliseet Indians, Michelle helped to develop a way of identifying the desirable type of muskrat root – *Acorus calamus* subspecies *americanus*. This subspecies is low in beta-asarone, a carcinogenic chemical. Sweet flag from tropical Asia and Europe has higher levels of this chemical, while introductions from North Asia are, like the North American subspecies, low in beta-asarone.

The second time I paid attention to sweet flag was when I tasted it in Kalmoes-fontein, South Africa, a place named after it – *kalmoes* being a Dutch word for sweet flag. Ade Badenhorst makes caperitif, a bottle of local flavour that, like Italian aperitifs, combines aromatic and bitter local herbs with wine. Each year a different batch is made that varies a little in relation to the year's abundance, yet a signature note is the heady aromatic and bitter flavour of sweet flag.

Sweet flag might also have a role in remedying the damage that humans cause to ecosystems. Heavy metal and polycyclic aromatic hydrocarbons (PAH) are contaminates generated by petroleum spills, metal mining and burning fossil fuels. They can accumulate in soil, and they have a negative effect on human health and food safety. But sweet flag, with its extensive roots, high biomass and adaptable nature, can grow in conditions where heavy metals and PAH are high, and

Sweet flag (*Acorus calamus*)

can remove these contaminants from soil by accumulating them in its tissues. Sweet flag can also flourish in sewage effluent, and remove excessive nutrients from the water, making it a plant that can help us clean up waterways and soil.

Where to see sweet flag
From spring to autumn, when its leaves are up, you can spot sweet flag at the edges of streams, ponds and lakes. Don't be tempted to dig up roots and taste the European form as it contains higher levels of carcinogenic beta-asarone. Foraging in other regions, you can taste it if you can tell the difference between the fertile diploid form and the infertile polyploid forms that are high in beta-asarone – an incentive to become botanically literate.

> Lough Neagh, Northern Ireland
> Wast Water, Cumbria, England
> River Itchen, Hampshire, England
> River Exe, Devon, England
> River Great Ouse, Cambridgeshire, England

Water shrews
Tiny but mighty
There are only a few venomous mammals in the world, and the water shrew (*Neomys fodiens*) is Britain's only one. Compounds in its saliva have paralytic effects – if you are a water shrimp, caddisfly or small frog.

Its diminutive teeth don't penetrate human skin, and the worst a water shrew can do to a person is give them a little rash. Their red teeth are not dripping in blood but coloured by iron deposits in the enamel of their tips, which strengthens them.

On land, the dainty feet and twirly tail stuck on a pudgy-looking body seem incongruous. Underwater, silvered with bubbles caught on their fur and streaking in pursuit of prey, water shrews are far more elegant. Their particularly dense fur that traps air is what makes it possible for them to dive and stay warm in winter, as it repels water and provides insulation, although in order to combat the buoyancy of trapped air, water shrews must work hard to dive down. Other hairy adaptations for their aquatic life are stiff hairs on their feet, which aid propulsion while swimming, and a keel of stiff hairs underneath their tail, which is used as a rudder. They detect prey using touch-sensitive whiskers on their snout.

With fifty per cent of their food being aquatic, water shrews' rapacious appetite makes them a barometer of water quality in the lowland areas they inhabit. They are associated with water low in nitrate pollution, where well-oxygenated water hosts a greater diversity of plants and animals. Their vulnerability to pollution is not only via the food they consume but also what they ingest when they groom their fur.

Water shrew (*Neomys fodiens*)

About nine centimetres long and weighing only up to eighteen grams, water shrews' body weight is a way to contextualise them in comparison to other species. In the course of a day, a water shrew needs to consume fifty per cent of its own body weight in food to survive. Male water shrews can have such hefty reproductive organs that they constitute a tenth of their entire bodyweight. Combined with having two or three litters of five to seven young in their year of life, water shrews seize the day with a brief but intensely lived existence, fuelled by hearts that beat more than 900 times a minute.

Where to see water shrews
Although widely scattered across mainland England and southern Scotland, water shrews are not commonly seen. They may be elusive even to ecologists, but people who spend a lot of time in and around freshwater habitats in central and eastern England at dawn and dusk have a decent chance of seeing one whizzing past on land or underwater. In winter, when vegetation is less dense, it can be easier to see them scurrying along the land edges of their territories.

> Stanborough Lakes, Hertfordshire, England
> Chimney Meadows, River Thames, Oxfordshire, England
> River Wensum, Norfolk, England
> Cannock Chase, Staffordshire, England
> River Stour, Bures, Suffolk/Essex, England

Yellow water-lilies

Traditional uses, future cure?

We were swimming in a river pool when my friend Cassy said, 'I've been testing that plant in my laboratory – can you take a photo of me with it?' Cassy works on anti-infective drug discovery and that plant was yellow water-lily (*Nuphar lutea*).

Cassy studies plants used by traditional medicine to treat infectious disease. She extracts their active components and tests if and how they work to inhibit microbial growth. Yellow water-lily's roots were used medicinally by the Iroquois, Menominee, Mi'kmaq, Ojibwa, Potawatomi and Sioux. It has also been a plant used as a food. Its fleshy roots are large and easily pulled out of the mud it grows in. In theory, this makes it easy to collect a large amount. But while the roots were eaten in the USA, in the UK people have found them unpalatable. Some botanists have started to separate the yellow water-lilies found in the USA and UK into separate species. Previously they were all called *Nuphar lutea*. Now some are being distinguished as different species – such as *Nuphar pumila* – due to their differentiation under molecular analysis, and others such as *Nuphar lutea* subsp. *advena* are being recognised as distinct subspecies. It seems the American species of yellow water-lily have more edible roots than the British yellow water-lily. Far more palatable are the seeds, which were consumed by Indigenous peoples in North America, and, as we know from archaeological records, were also cooked and eaten in Europe.

It is an interesting juxtaposition that a plant encountered when swimming in ponds and slow-moving rivers could become a vital defence for us in the rapidly evolving arms race between humans and bacteria. The World Health Organization considers antibiotic resistance to be one of the biggest threats to human health. With their stumpy petals, yellow water-lilies are not the most elegant of water-lilies, but they merit appreciation for feeding our ancestors and offering hope for our future.

It is worth pausing and quietly looking at yellow water-lilies to see the other animals that make use of them. Frogs can perch on the leaves and rest while remaining camouflaged from predators, and blue damselflies often lay their eggs attached to the underside of yellow water-lily leaves. Like many plants, yellow water-lilies have more than one form of leaf; their submerged leaves look like large, crumpled lettuce leaves. These unstructured leaves are able to move and yield to shifting currents in the water.

Where to see yellow water-lilies
Distributed across Eurasia and North America, yellow water-lilies are found in sheltered areas of lakes, and languid rivers. They don't inhabit water where they would be buffeted by wind or fast-moving currents. As they grow with their roots in mud and their leaves and flowers floating on the water surface, they tend to be at the shallow margins of lakes and rivers, so be prepared for a muddy entry. It is worth ducking under to see light coming through their leaves, and flipping the leaves over to see if anything has laid eggs on them. They flower from June to August.

> Warleigh Weir, Somerset, England
> Lough Leane, County Kerry, Ireland
> River Thames, Berkshire, England
> Stroan Loch, Dumfries and Galloway, Scotland
> Llangorse Lake, Powys, Wales

Yellow water-lilies (*Nuphar lutea*) >

Illustrations

Wildlife that savours reed beds (*page x*)

1 Swallowtail butterfly (*Papilio machaon britannicus*)

2 Reeds (*Phragmites australis*)

3 Bearded tit (*Panurus biarmicus*)

4 Milk parsley (*Peucedanum palustre*)

5 Swallowtail caterpillar (*Papilio machaon britannicus*)

Jellyfish found around Britain and Ireland (*page 21*)

1 By-the-wind sailors (*Velalla velalla*)

2 Barrel jellyfish (*Rhizostoma pulmo*)

3 Portuguese man-of-war (*Physalia physalis*)

4 Moon jellyfish (*Aurelia aurita*)

5 Mauve stinger (*Pelagia noctiluca*)

6 Blue jellyfish (*Cyanea lamarckii*)

7 Compass jellyfish (*Chrysaora hysoscella*)

8 Lion's mane jellyfish (*Cyanea capillata*)

Sharks that inhabit British and Irish seas (*page 18*)

1 Frilled shark (*Chlamydoselachus anguineus*)

2 Bramble shark (*Echinorhinus brucus*)

3 Basking shark (*Cetorhinus maximus*)

4 Shortfin mako shark (*Isurus oxyrinchus*)

5 Porbeagle shark (*Lamna nasus*)

6 Kitefin shark (*Dalatias licha*)

7 Nursehound shark (*Scyliorhinus stellaris*)

8 Starry smooth-hound shark (*Mustelus asterias*)

9 Thresher shark (*Alopias vulpinus*)

10 Angel shark (*Squatina squatina*)

11 Bluntnose sixgill shark (*Hexanchus griseus*)

12 Small-spotted catshark (*Scyliorhinus canicula*)

13 Angular roughshark (*Oxynotus centrina*)

14 Smooth hammerhead shark (*Sphyrna zygaena*)

15 Tope shark (*Galeorhinus galeus*)

16 Sharpnose sevengill shark (*Heptranchias perlo*)

17 Greenland shark (*Somniosus microcephalus*)

18 Blue shark (*Prionace glauca*)

Seaweed diversity (*page 84*)

1 Sugar kelp (*Saccharina latissima*)

2 Dulse (*Palmaria palmata*)

3 Pepper dulse (*Osmundea pinnatifida*)

4 Twisted wrack (*Fucus spiralis*)

5 Gutweed (*Ulva intestinalis*)

6 Oarweed (*Laminaria digitata*)

7 Egg wrack (*Ascophyllum nodosum*)

8 Irish moss (*Chondrus crispus*)

9 Sea lettuce (*Ulva lactuca*)

10 Bladder wrack (*Fucus vesiculosus*)

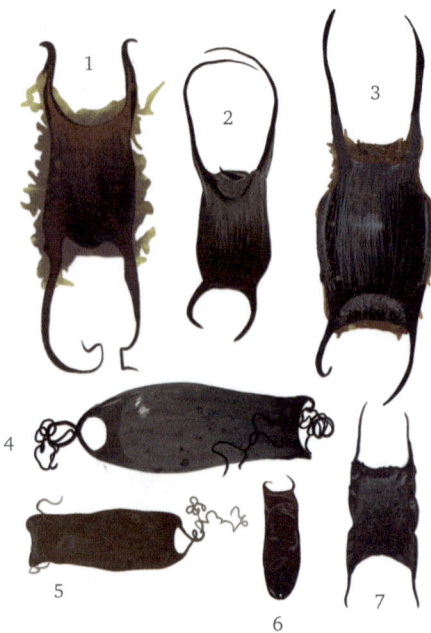

Shark and egg cases (*page 98*)

1 Undulate ray (*Raja undulata*)

2 Cuckoo ray (*Leucoraja naevus*)

3 Blonde ray (*Raja brachyuran*)

4 Nursehound (*Scyliorhinus stellaris*)

5 Small spotted catshark (*Scyliorhinus canicula*)

6 Blackmouth catshark (*Galeus melastomus*)

7 Starry skate (*Amblyraja radiata*)

Going Further

Not just useful references if you want to dive deep into identifying particular groups of plants and animals, but also some pointers towards groups that can help you to go further.

This isn't a comprehensive directory. I have yet to find a book or website covering all the fish and squelchy critters (jellyfish, nudibranchs, etc.) that catch my attention. While there are regional and group-specific guides that I find useful, they are rather niche to inventory here.

Accessibility

Reaching wild waters isn't as straight-forward for everyone as it should be. People can help navigate whatever an individual's obstacles are. Certainly going on hikes and swims with my friend Cassy, who is an amputee, hasn't restricted our fun. Directly knowing the terrain or asking people who have been to our proposed destination has helped with planning timing and following paths that are less steep (it's hard going uphill with one leg doing all the work!). We have got better at stashing her leg for swims and drifting on rivers – a large tow float is good for keeping it dry, sand-free and with us.

If you don't have 'outdoorsy' friends already, there are lots of supportive collectives who can be found online and joined outdoors. Some examples are:

> Able 2 Adventure
> *www.able2adventure.co.uk*
> Black Girls Hike
> *www.bghuk.com*
> Boots and Beards
> *www.bootsandbeards.co.uk*
> Land in our Names
> *www.landinournames.community*
> Steppers UK
> *www.instagram.com/steppers_uk*
> We Go Outside Too
> *www.instagram.com/wegooutsidetoo*

Whatever your access barrier, there is probably a local or national group that can help you. Try searching online, and asking people. It can take a bit of chasing to get to the person who can signpost you towards what you need, but friendly help is out there.

At local level, councils are getting on board and facilitating access to the outdoors with specific schemes such as provision of beach wheelchairs and rugged wheelchairs for borrowing. Charities and other organisations that manage outdoor spaces are also beginning to work on improving access. For example, the Canal & River Trust have an accessibility map intended to help everyone navigate these waterways:

> *www.canalrivertrust.org.uk/enjoy-the-waterways/walking/accessibility-map*

Canal paths, being intended for horses to plod along, are flat and convenient for wheelchairs and prams, but some of the access points are via stairs so it's worth checking online before trying to reach the path.

Seaweeds

When I heard Dr Christine A. Maggs speak about seaweeds, she made getting to grips with different seaweed groups and whittling down to individual species easy to understand. She also highlighted dynamic aspects of seaweed distribution and novel uses. For identification purposes, you can get the book she co-authored with several other seaweed experts:

> *Seaweeds of Britain and Ireland* (second edition, 2017), by Francis St.P.D. Bunker, Juliet A. Brodie, Christine A. Maggs and Anne R. Bunker.

It's remarkable how much information is crammed into the book. Alongside photographs of seaweeds as they are growing underwater, there are distribution maps and scale drawings. This is accompanied by clearly laid-out text descriptions.

Plants

The Wild Flower Key: How to identify wild flowers, trees and shrubs in Britain and Ireland (2006) by Francis Rose, revised and updated by Clare O'Reilly, is a user-friendly and comprehensive identification guide for plants in Britain and Ireland. You can use the 'key' – a series of questions and answers that guide you towards identifying a plant. Or you might browse the illustrations to find similar-looking plants and then read the description text to check if what you see matches the identifying features described for the species. This is a very practical book; make use of the ruler on the back cover for checking flower and leaf measurements. It is also written in English that is as plain as possible, and there is a glossary at the back for the indispensable botanical terms.

Birds

The RSPB has a comprehensive online guide:

> *www.rspb.org.uk/birds-and-wildlife/wildlife-guides/bird-a-z*

Though driven by bird conservation, a lot of the sites managed by the RSPB are also havens for other watery wildlife. Do have a look at the Reserves A–Z on their website, as you may well enjoy visiting these places.

Insects

In terms of number of species and numbers of individuals within each species, insects should be one of the most significant groups of wildlife, but they are often neglected by our attention. Buglife works on reminding us that these creatures can be charismatic as well as important within ecosystems. Use their online identification guide for help working out what you have seen, or enjoy browsing their bug directory:

> *www.buglife.org.uk/bugs/bug-directory*